William A. Moale, Henry N. Martin

A Handbook of Vertebrate Dissection

Volume 2

William A. Moale, Henry N. Martin

A Handbook of Vertebrate Dissection
Volume 2

ISBN/EAN: 9783337255701

Printed in Europe, USA, Canada, Australia, Japan

Cover: Foto ©berggeist007 / pixelio.de

More available books at **www.hansebooks.com**

HANDBOOK

OF

VERTEBRATE DISSECTION.

BY

H. NEWELL MARTIN, D.Sc., M.D., M.A.,

PROFESSOR IN THE JOHNS HOPKINS UNIVERSITY,

AND

WILLIAM A. MOALE, M.D.

PART II.

HOW TO DISSECT A BIRD.

NEW YORK:
MACMILLAN AND CO.
1883.

Trow's
Printing and Bookbinding Company
201 *to* 213 *East Twelfth Street*
NEW YORK

PREFACE.

THE object of the Handbook of Vertebrate Dissection was stated so fully in the Preface to Part I., that it is unnecessary to go over the same ground again. To prevent disappointment it may, however, be well to state that our intention is not to enable a student to determine species, but to give the young morphologist practical directions assisting him to learn for himself what a fish, an amphibian, a reptile, a bird, and a mammal are, when considered from an anatomical point of view and contrasted with one another. Specific and generic characters are therefore hardly touched upon. The collector who desires to find out the name of any given bird, must seek the diagnostic characters of the various species elsewhere; so far as American birds are concerned, this classificatory work has been so well done by Coues in his Key that it would be presumptuous to attempt to improve it. What we have sought is to give such directions as will enable the student who follows them to have a good knowledge of the anatomical characters of the Birds as a group of vertebrates, paying little heed to the minor differences which exist between different birds. Accordingly, many points of structure not particularly avian have been treated with much less detail than those which are ; the bird's skeleton is so very characteristic that it has been treated in considerable detail, while, on the other hand, only such muscles have been described as present peculiar characters in all or mos birds; and so throughout.

It is due to Dr. Moale to state that almost the whole work of preparing this volume and seeing it through the press has fallen upon him. He is therefore entitled to far the greater proportion of any gratitude toward us which may be felt by those who, we trust, will find the book useful.

H. NEWELL MARTIN.

BALTIMORE, October, 1882.

CONTENTS.

	PAGE
ZOÖLOGICAL POSITION OF THE DOMESTIC PIGEON,	89
ANATOMY OF THE DOMESTIC PIGEON,	93
INDEX,	169

DESCRIPTION OF THE FIGURES.

FIG. 1 represents the roof of the skull; FIG. 2, its base; FIG. 3, the skull as seen from the left side; and FIG. 4, the inner side of the left half; FIG. 5 is a diagram showing the relative position of the bones of the skull; FIG. 6 represents the bones of the leg and foot as viewed from the inner side.

The references in the figures of the skull are as follows: 1, the lachrymal; 2, the palatine; 3, the pteregoid; 4, the quadrate; 5, the maxilla; 6, the jugal; 7, the quadrato-jugal; 8, the premaxilla; 9, the nasal; 10, the ethmoid; 11, the interorbital septum; 12, the frontal; 13, the squamosal; 14, the parietal; 15, the periotic capsule; 16, the basi-sphenoid; 17, the ali-sphenoid; 18, the supra-occipital; 19, the exoccipital; 20, the basi-occipital; 21, the foramen magnum; 22, the occipital condyle; 23, the external auditory meatus; 24, the rostrum of the basi-sphenoid; 25, exit of optic nerves; 26, opening of Eustachian tubes; 27, pituitary fossa; 28, the carotid foramen; 29, exit of the seventh nerve; 30, foramen for the ninth, tenth, and eleventh nerves; 31, exoccipital foramen for the twelfth nerve; 32, foramen for the eighth nerve.

The references in FIG. 6 are as follows: 1, the fibula; 2, the cnemial process of the tibia; 3, the proximal segment of the tarsus anchylosed to the distal extremity of the tibia; 4, the distal segment of the tarsus anchylosed to the proximal extremities of the united metatarsal bones; 5, the metatarsal bone of the hallux; 6 and 7, the distal extremities of the second and third metatarsals; 8, the phalanges.

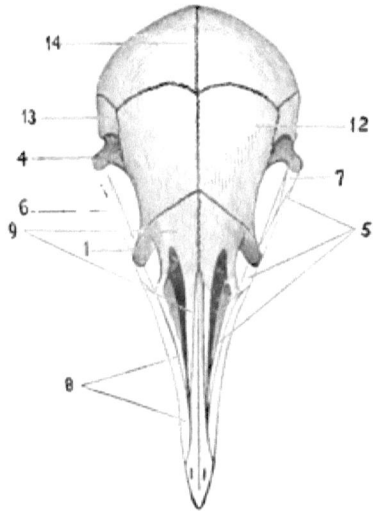

FIG. 1.

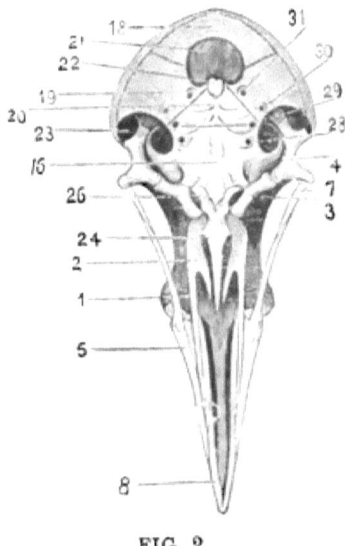

FIG. 2.

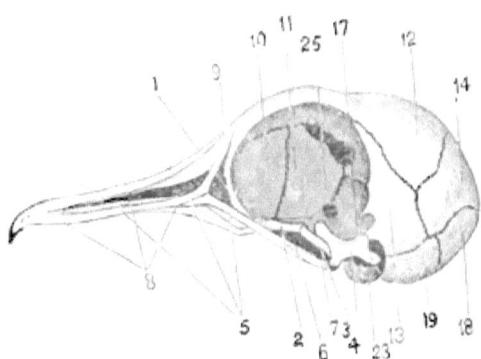

FIG. 3.

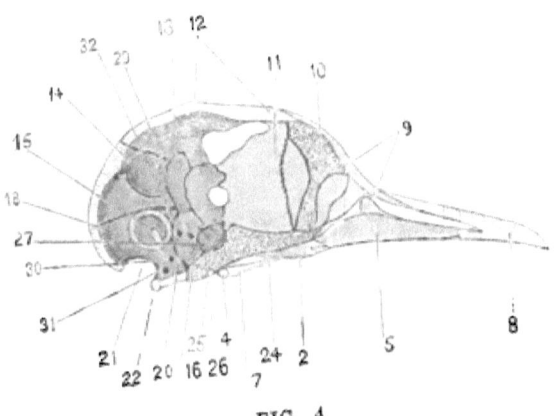

FIG. 4.

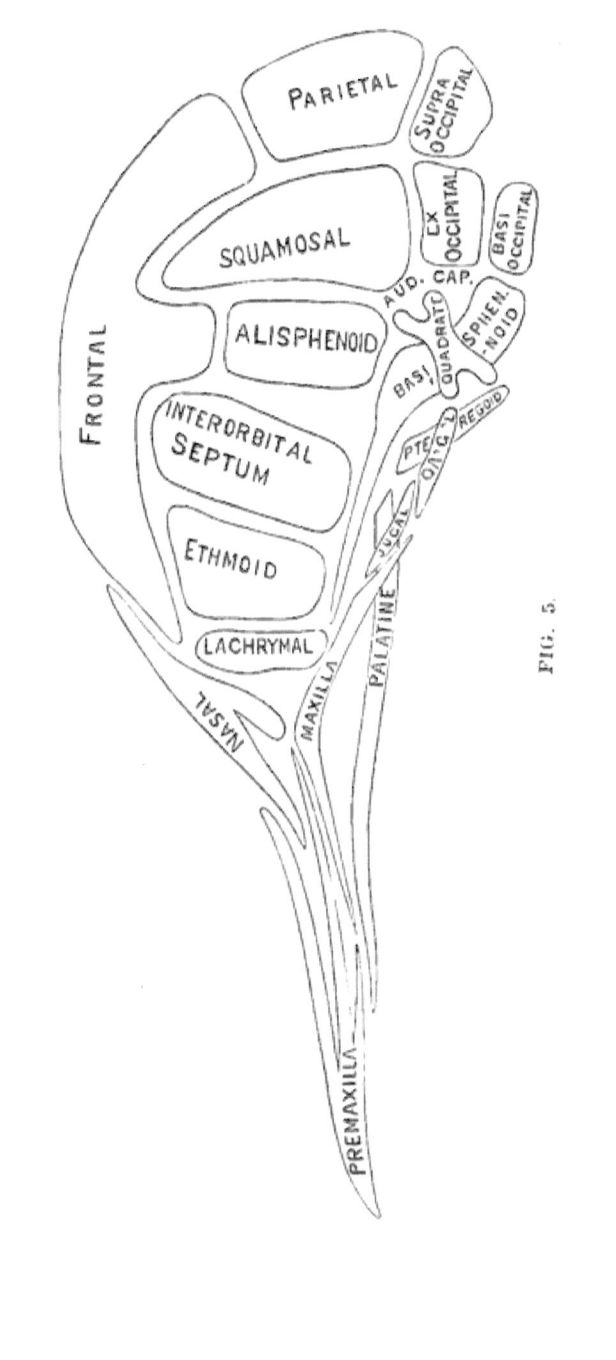

Fig. 5.

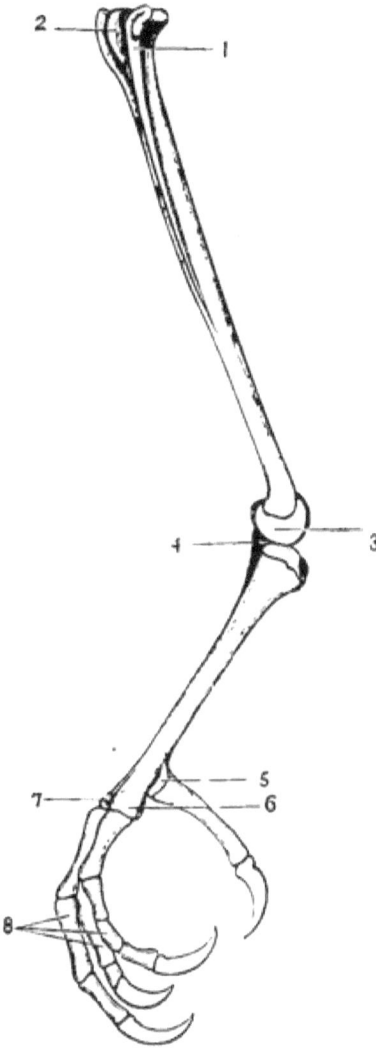

FIG. 6.

ZOOLOGICAL POSITION OF THE DOMESTIC PIGEON.

SUBKINGDOM, **Vertebrata.**—DIVISION, **Sauropsida.**—CLASS, **Aves.**—ORDER, **Carinatæ.**—SUBORDER, **Schizognathæ.**—FAMILY, **Peristeromorphæ.**—GENUS, **Columba.**—SPECIES, **livia.**—VARIETY, **domestica.**

Characters of the Sauropsida.*

1. Almost always an epidermic skeleton, in the form of scales or feathers.
2. The vertebral centra are ossified, but have no terminal epiphyses.
3. The skull has a completely ossified occipital segment and a large basi-sphenoid. There is no separate para-sphenoid in the adult. The proötic is always ossified, and either remains distinct from the opisthotic and epiotic during life, or unites with them only after they have anchylosed with adjacent bones.
4. There is always a single convex occipital condyle, into which the ossified ex-occipitals and basi-occipital enter in various proportions.
5. A mandible is always present, and each ramus consists of an articular ossification and several membrane-bones. The articular is connected with the rest of the skull by an ossified quadrate.

---

* The characters given are taken, with slight modification, from Huxley.

6. The apparent ankle-joint lies between the proximal and distal divisions of the tarsus.

7. The alimentary canal terminates in a cloaca.

8. The heart is trilocular or quadrilocular; some of the blood-corpuscles are red, oval, and nucleated.

9. The aortic arches may be two or more; when only one persists in the adult it is on the right side.

10. Respiration is never performed by branchiæ; and after birth always by lungs, in which the bronchi do not branch dichotomously.

11. A thoracic diaphragm may exist, but never forms a complete partition between the thoracic and abdominal viscera.

12. The cerebral hemispheres are never united by a corpus callosum.

13. The reproductive organs open into the cloaca. The oviduct is a Fallopian tube with a posterior uterine dilatation.

14. There are no mammary glands.

15. All are oviparous or ovo-viviparous.

16. The Wolffian bodies are replaced functionally by permanent kidneys.

17. The embryo has an amnion and a large respiratory allantois; and develops at the expense of a large mass of food stored up in the egg.

Characters of **Aves**, as distinguished from Reptilia, which with them constitute the division Sauropsida.

1. The exoskeleton consists mainly of feathers. Ossifications of the dermis are rare, and never take the form of bony plates.

2. In all recent birds, the centra of the cervical vertebræ, at least, have subcylindrical articular faces. If, as in some birds, the faces of the centra of the other vertebræ are spheroidal, they are opisthocœlus, which is the rarest arrangement among reptiles.

3. The proper sacral vertebræ of birds—that is to say, those between, or through, the arches of which the roots of the sacral plexus pass—have no expanded ribs abutting on the ilia.

4. The sternum has no costiferous median backward prolongation, all the ribs being attached to its sides. The cartilaginous sternum is replaced, in the adult, by membrane-bone, and ossifies from two to five, or more, centres.

5. When an interclavicle exists, it is confluent with the clavicles.

6. The manus possesses not more than three digits, and not more than the two radial digits have claws.

7. The ilia are greatly prolonged in front of the acetabulum, the inner wall of which is membranous. The pubes and ischia are directed backward, more or less parallel with one another, and the ischia never meet in a ventral symphysis.

8. The astragalus sends up a process on to the front of the tibia, and early anchyloses with the latter bone. In this character, Birds differ from all existing Reptiles. The foot contains not more than four digits. The first metatarsal is, almost always, free, shorter than the rest, and incomplete above. The other three are anchylosed together, and with the distal tarsal bone, to form a tarsometatarsus.

9. Only one aortic arch, the right, is present. Only

one arterial trunk, the pulmonic, is given off from the right ventricle. The arterial and venous currents communicate only through the capillaries.

10. The blood is hot. There are three semi-lunar valves at the origin of the aortic and pulmonary trunks. In all existing birds the extremities of the chief pulmonary passages terminate in air-sacs.

11. The corpora bigemina are thrown down to the sides and base of the brain.

Characters of the Schizognathæ.

1. The metacarpals are anchylosed together. The tail is considerably shorter than the body.

2. The sternum is provided with a keel.

3. The vomer is narrow behind; the pterygoids and palatines articulating largely with the basi-sphenoidal rostrum.

4. The maxillo-palatines free.

5. The vomer, when present, pointed in front.

THE ANATOMY OF THE DOMESTIC PIGEON
(*Columba livia, var. domestica*).

1. General External Appearance.*

a. Note the general spindle form of the head and trunk of the bird, tapering gently to each end, and enabling it to cleave the air in flying, while diminishing the drag behind due to rarefaction of the atmosphere.

b. Note the main divisions of the body; head, neck, trunk, and limbs; the peculiar modification of the fore limbs to form *wings*.

c. The *feathers* covering most of the body; their absence on eyelids, beak, and the lower parts of the legs.

2. On the Head, study —

a. The conical *bill*, which varies much in length in different specimens; the mouth-opening lies between the upper and lower divisions (*mandibles* of ornithology) of the bill.

b. Each so-called mandible is hard and horny at its tip, but becomes softer near the angle of the gape.

c. Posteriorly the upper mandible presents two

---

* The different regions of the bird's body have been mapped out very minutely by ornithologists and given special names. For these consult Coues, Key to North American Birds.

soft, skinny swellings, beneath each of which is the narrow, elongated opening of a nostril.

- d. Note that the tip of the upper mandible overlaps the lower; the bill, therefore, belongs to the type called *epignathous*.
- e. Holding the head firmly between the thumb and finger of one hand, with the other seize the upper mandible and note that it possesses some mobility; thus differing from the rigidly fixed upper jaw of a mammal.
- f. The circular *eye*, with its bright orange *iris*. The upper and lower *eyelids* tolerably opaque; the translucent, *nictitating membrane* attached at the inner corner of the exposed part of the eye, but capable of being drawn back so as to completely cover it.
- g. The *external auditory meatus* may be found behind and below the eye on pushing aside some peculiar feathers which cover it. These feathers have loose vanes, and so, while covering and protecting the ear, oppose but little obstacle to the passage of sound-waves.

3. **Spread out the** Wings. Note—
    - a. Their great size, due mainly to the stiff feathers which form most of their expanse.
    - b. The concavity of the spread-out wing below, and the slight convexity of its upper surface.
    - c. The main divisions of the limb, apart from the feathers, into *arm*, *forearm*, and *manus*.

4. **On the Hind Limbs**, note—
    - a. The *thigh*, *crus*, and *foot*. The latter consists of an elongated, unfeathered *tarso-metatarsus*,

to which are articulated four digits (forming the *pes* of ornithology, though in comparative anatomy, the tarso-metatarsus is properly a part of the *pes*); one of the toes is turned backward, three forward.

 b. The curved, pointed *claws* on the ends of the digits.

 c. The red, rectangular, horny scales (*scutellæ*) on the anterior aspect of the tarso-metatarsus and the upper sides of the toes.

5. The **Structure of a Typical Feather.**—The horny covering of the bill, the beak, the scutellæ, and the claws constitute, with the feathers, the epidermic exoskeleton of the bird. The feathers being characteristically avian, must be studied in more detail.

6. Pluck one of the large feathers from the wing. Note—

 a. It possesses a main stem or *scapus* composed of quite different proximal and distal portions.

 b. The proximal part (quill or *calamus*) is cylindrical, translucent, and hollow. Its proximal end, normally embedded in a follicle of the skin, contains a soft, reddish vascular pulp, and presents a terminal opening (*inferior umbilicus*). The distal portion of the calamus contains a number of dry scales.

 c. The remaining part of the scapus, called the *rachis*, is somewhat rectangular in cross section, and tapers distally to a point. It is whitish in color, opaque, and filled with a dry soft pith.

d. Where rachis joins calamus there will be found on the under side of the feather a depression in the scapus, fringed by slender, feathery filaments, and presenting within it a small aperture (*superior umbilicus*), which leads into the cavity of the calamus.

e. Attached to two opposite sides of the rachis are the *vanes*, or *vexillæ*, the flat, expanded portions of the feather. In some feathers the two vanes are of approximately equal size; in others they are very unequal. Each vane will be found to be made up of a large number of separate portions, which, however, cohere pretty firmly, and make the whole vane a continuous resisting membrane.

f. The primary constituents of the vane are the *barbs* (*barbæ*). Each barb is a flattish, triangular plate, with its surfaces turned toward the barbs which precede and succeed it, and its base attached to the rachis. The barbs are not set on vertically to the long axis of the rachis, but obliquely, with their free ends sloping toward the tip of the feather.

g. Examine a portion of the vane with an objective magnifying about thirty diameters. From each side of each barb a number of thin plates, the *barbules*, will be seen to arise, and cross obliquely the barbules of the neighboring barbs.

h. Isolate a barb and examine with the same magnifying power as above. The barbules on its proximal side (that nearer the calamus) will be seen to be closely packed, and to end in slender filaments (*barbicels*). The barbules on its distal side are looser and end in larger filaments, which bear fine processes and knobs. Normally these knobbed threads lie over and hook upon the proximal barbules of the succeeding barb, and so bind the whole into a coherent membrane adapted to oppose a firm resistance to the air.

7. Varieties of Feathers.

a. Such a feather as that just described is known as a *penna*, or, as such feathers determine the general form of the birds, as a *contour feather*.

b. Lying beneath the contour feathers are *down feathers* (*plumulæ*), not very abundant on the pigeon. Each has a short, weak scapus, with long, soft barbs, whose barbules do not interlock so as to form a compact vane. The whole, therefore, forms a little, downy tuft.

c. Pluck one side of the belly of the bird ; among the other feathers will be found a number of *filo-plumæ*, or hair feathers. They possess a slender stem, not clearly marked off into calamus and rachis, and a few barbs near the distal end, not interlocked so as to form a firm vane.

d. The *semi-plumæ* may be found in great numbers about the under surface of the body. Their scapus is like that of a penna, but the vane is loose and downy, its barbules not being interlocked.

I*

8. The Arrangement of the Feathers.

a. Certain parts of the body (eyelids, bill, tarso-metatarsus and pes) are obviously unfeathered. Closer examination shows, further, that many parts of the body which are overlapped by feathers really give origin to none. To observe the tracts of the skin (*pterylæ*) from which feathers arise, and those (*apteria*) which give origin to no feathers, it is necessary either to clip closely with scissors the plumage (technically, *ptilosis*) of an adult pigeon, or, better still, to obtain and examine a young pigeon (squab) after its feathers have appeared, but before it is able to fly. On either specimen it may be readily observed that the feathers, particularly the contour feathers, are borne only on limited tracts ; and that the feather-bearing tracts (*pterylæ*) are separated by bare spaces.*

9. The Wing Feathers.

a. Pluck one wing of your pigeon, and make out the divisions and bones (12, *g ;* 40–45) of the fore limb. Then on the other side note the insertion, size, number, and arrangement of the following special groups of feathers.

b. The *remiges*, to which the wing mainly owes its extent and form. They are pennæ arranged along the posterior margin of manus and forearm, and fall into two groups, the *primaries* and *secondaries*.

---

* For the technical names of the different pterylæ consult Coues, Key to North American Birds, p. 5.

ARRANGEMENT OF FEATHERS.

c. The *primaries* are ten in number and are inserted upon the manus.

d. The *secondaries* are thirteen in number and arise from the antebrachium.

e. The *wing coverts* (*tectrices alæ*) are the feathers which overlap the bases of the remiges; according as they lie on the upper or under side of the wing they are known as the upper or lower coverts (*tectrices superiores* or *inferiores*).

f. The *upper* **wing coverts** are divided into *primaries*, which spring from the manus, and *secondaries*, which mostly arise from the forearm. The secondaries are arranged in three **overlapping rows**, known respectively as the *greater, median,* and *lesser* **coverts**. The greater **coverts** are the longest and overlap the secondary remiges; the median are the next row, not very **well marked in the** pigeon; the lesser coverts include all the remaining secondary upper coverts.

g. The lower wing coverts are arranged much like the upper, but as they all pretty closely resemble one another, they are not described in separate groups.

h. The *alula*, or the *bastard quills*, is the name given to a distinct tuft of feathers which spring from the thumb or pollex.

10. The Tail Feathers.

a. Answering to the remiges of the wing are twelve large, stiff feathers, forming the main expanse of the tail; they are called the *rectrices*. If the tail is spread out it will be seen that these feathers are so arranged that the median pair lie above or dorsal to all the rest, and each of the remaining feathers overlaps the feather on its outer side, and is overlapped by that on its inner side.

b. The *tail coverts* are pennæ which overlap and underlie the rectrices; they naturally fall into two groups—the upper and lower tail coverts respectively.

11. Finish plucking the pigeon. Note on the skin the elevations where the feathers were pulled out, each presenting a minute aperture which leads into the follicle in which a feather was inserted. Between the regions of the skin (pterylæ) presenting these elevations will be seen smoother tracts (the apteria) devoid of them.

12. On the plucked bird, note—

a. The head, tapering in front to the bill, and rounded behind; on it the opening of the external auditory meatus.

b. The long and flexible neck, of sufficient length to allow the beak to reach the uropygial gland (12, *e*) on the rump.

c. At the base of the neck, on its ventral aspect, the swelling due to the *crop*, which nearly always contains some hard grains.

d. The somewhat ovoid trunk, ending posteriorly in an enlargement, the *uropygium*, on which the tail feathers were inserted.

e. On the dorsal surface of the uropygium, in the middle line, is a conical elevation, on

.which the duct of the uropygial gland opens. On compressing the elevation a drop of oil can be squeezed out of the gland.

f. On the ventral side of the uropygium, at its proximal end, note the cloacal aperture; a transverse opening, with puckered, elevated margins.

g. Note the main divisions of the fore limb: brachium, antebrachium, and manus. In the manus observe the short, separate, pointed *pollex* (thumb). The two other digits of the hand (second and third) may be felt through the skin and other soft parts which cover them.

h. Note on the hind limb its main divisions: femur, crus, and pes; the latter made up of the tarso-metatarsus and the pes of ornithology.

13. Opening the mouth, note—

a. The two horizontal, fleshy plates, with denticulated edges, meeting one another in the middle line, and forming most of the roof of the mouth.

b. Separating these plates, examine the nasal cavity above them.

c. Note that this cavity is not subdivided into two nostril chambers, as the rostrum of the sphenoid, which, covered by mucous membrane, projects on its roof, does not reach its floor.

d. Where the palatal membranes diverge behind, the nose communicates with the pharynx by a single posterior nostril.

e. The horny, hastate tongue, only attached by a small part of its under surface.

f. Open the bill widely and draw the tongue forward with forceps; behind it will be seen, on the ventral aspect of the pharynx, the slit-like opening of the larynx (*aditus laryngis*).

g. Beyond the laryngeal opening note the transverse fold of membrane with a denticulated edge projecting from the floor of the pharynx, and the similar larger membrane hanging down from its roof.

14. The Bony and Cartilaginous Skeleton.—For the satisfactory study of the skeleton the student should have beside him for reference a prepared articulated specimen, and should himself disarticulate and examine another specimen. For this purpose boil the pigeon thoroughly, and then carefully pick the bones clean.

15. On the articulated specimen note the general arrangement of the skeleton; its spinal column and skull, ribs and huge sternum; limb arches and limbs.

16. The Skull.—Before proceeding to clean the skull, note on the prepared specimen the position of various slender, bony plates and bars belonging to it. Take great care not to injure these while removing the soft parts. If possible leave also the membranous part of the interorbital septum.

17. Observe the general form of the cranium, with its curved roof, flatter base, rounded posterior, and pointed anterior end. On holding the skull up to the light, the large air-cells in many of its bones may be seen. Make outline drawings of it as seen from above, below, behind, and laterally.

18. In adults very many bones of the skull are inseparably anchylosed together and their lines of union not visible. The structure of the skull is therefore much facilitated by having that of a squab (in which the anchylosis is less complete) to compare with the skull of the adult.

19. Remove the lower jaw, and **on the base of the skull** note—

- *a.* In front the long, narrow interspace between the *palatine processes* of the maxillæ (21, *c*).
- *b.* Further back are the slender palatines (21, *b*), with a larger opening, the *posterior naris*, between them.
- *c.* Dorsal to the palate bones and between them is the conical pointed *rostrum* of the *basi-sphenoid* (21, *p*).
- *d.* The *pterygoids* diverge from the posterior extremities of the palatines and articulate at their outer ends with—
- *e.* The *quadrates*, projecting ventrally and forward and bearing the articular facets for the mandible.
- *f.* Behind the quadrate and rather on the side than the base of the skull the opening of the *external auditory meatus;* on its floor may be seen several apertures, the most anterior and largest being the commencement of the *Eustachian canal*, and a smaller one immediately

behind it, the opening of the canal through which the *columella auris* passes (158, *f*) on its way to the fenestra ovalis. The other openings communicate with the air-cells of the side and base of the skull.

- *g.* The columella may be seen *in situ* if the skull has been carefully cleaned.
- *h.* In the median line behind is the *foramen magnum*, and immediately in front of it the *single occipital condyle*.
- *i.* Just behind the junction of its rostrum with the basi-sphenoid, in the median line, note the common *opening of the Eustachian tubes*.
- *j.* Near the lower margin of the posterior part of the orbito-temporal fossa the *foramen ovale* for the fifth nerve (140) may be seen, situated about the inferior margin of the alisphenoid (21, *q*).
- *k.* On the median side of the lower margin of the external auditory meatus the foramen, through which the ninth, tenth, and eleventh (145, *a*; 146) nerves pass out.
- *l.* In front of this foramen is an opening in which may be found the foramen for the seventh nerve, and also, larger than this and anterior to it, the aperture of the carotid canal.
- *m.* The foramen for the twelfth is immediately external to the occipital condyle.

20. Note on the skull viewed laterally—
    - *a.* The large *orbito-temporal fossa*, the floor of which is almost entirely wanting.
    - *b.* The *bony bar*, composed of jugal and quadrato-

THE SKULL.

jugal beneath it, extending from the distal end of the quadrate (19, *e*) to the maxilla.

- *c.* Anterior to the orbito-temporal fossa the triangular *lachrymo-nasal opening.*
- *d.* On the upper outer surface of the beak the elongated opening of the *anterior nares.*
- *e.* The *interorbital septum* partly membranous in its central portion.
- *f.* Observe that the upper portion of the posterior orbital wall is membranous, leaving an opening in the macerated skeleton, which leads into the cranial cavity.
- *g.* Note that the beak is united to the skull only through the medium of the nasal bones.

21. Boil the skull again and remove the bones in the following order as they become loose. Each bone should be sketched into the outline drawings, already prepared (17), as it is removed, so as to show as clearly as possible its various relations.

- *a.* At the anterior end of the orbit the *lachrymal bone*, which, viewed from its orbital surface, is flattened, concave from above downward, and has somewhat the shape of an elongated S. Remove this, and note the process which projects forward from the upper half of its outer border, partially closing in the *lachrymo-nasal space* seen between the nasal and maxillary bones, and helping to bound the nasal cavity externally.
- *b.* On the base of the skull the long, slender *palatines*, articulating, by the anterior two-thirds of their upper surface, with the maxillary

processes of the premaxillæ. Posteriorly they are considerably enlarged, and movably articulated internally with the rostrum of the basi-sphenoid on which they slide; also by their posterior extremities with the pterygoids.

c. Note the *pterygoids*, a pair of short, bony bars articulating movably with the palatines and rostrum of the basi-sphenoid, and thence directed outward and backward to articulate with the inner side of the inner condyloid processes of the quadrate bones. About the middle of their posterior borders the pterygoids articulate with a pair of processes from the basi-sphenoids.

d. The *quadrate* is irregular, being constricted in the middle; it expands above and below into a double articular extremity. Its upper end articulates movably with the squamosal, alisphenoid, and exoccipital. Its lower end articulates with the mandible, quadrato-jugal, and pteregoid.

e. The *maxilla* consists of two portions: a posterior, slender and rod-like, united by a squamous suture to the jugal and quadrato-jugal, and an anterior, meeting the former at an obtuse angle, flattened and grooved above to be applied to the inferior surface of the maxillary process of the premaxilla; it sends inward and downward a palate process, which does not, however, meet its fellow of the opposite side in the middle line, an interval being left between them.

f. The *jugal* forms part of the middle of the

suborbital bony bar, and is joined in front to the maxilla, behind to the quadrato-jugal. It is a mere scale of bone.

g. The *quadrato-jugal* constitutes the posterior portion of the same bar, and is united in front to the jugal; behind it articulates with a facet upon the outer aspect of the more external of the two divisions of the distal end of the quadrate.

h. The *premaxilla* is large, and forms, with its fellow of the opposite side, the greater portion of the bone of the beak. It is composed of three parts united at the tip of the upper mandible; these parts are the *maxillary* and *palatine processes* (the latter small), uniting respectively with the maxillary and palatine bones by squamous sutures; and the *ascending* or *nasal process* which articulates on its inner side with its fellow, and externally with a process from the nasal bone of the same side.

i. The *nasals* meet along the median line. Posteriorly they are expanded and flattened, and lie on the ethmoid; anteriorly they are forked, the inner division meeting the ascending process of the premaxilla, and the outer articulating with the maxilla, near its angle.

j. The *ethmoid* is a perpendicular plate of bone much expanded above where it underlies the nasals; it has projecting from its sides a pair of lateral masses which represent the prefrontals. The ethmoid forms the septum between the nasal cavities behind, and also part of the inter-orbital septum. It articulates above with the frontals and nasals, below with the rostrum of the basi-sphenoid, and externally with the lachrymals. Behind it is continuous with the inter-orbital septum, the posterior porton of which probably represents the *presphenoid*.

k. After removing the ethmoid note the *rod of cartilage* which lies beneath its lower margin in the groove on the upper surface of the rostrum of the basi-sphenoid; this is the *naso-ethmoidal cartilage*, which is continuous in front with the cartilaginous inter-nasal septum, and behind through a canal in the anterior vertical portion of the basi-sphenoid, with a plate of cartilage which overlies the surface of the sella turcica (21, *p*).

l. The *frontals*, not separable in the adult, form the roof of the orbits, and behind, where they are expanded, cover in the anterior portion of the brain-case. They send downward large processes which form part of the posterior orbital walls. In front they articulate with the nasals, ethmoid, and lachrymals; behind with the parietals; externally with the squamosals, and internally with each other. Their orbital processes anchylose with the alisphenoids and squamosals.

m. The *squamosal* is a thin, flat bone on the side of the cranium, just behind the orbit. It articulates with below and overlaps the periotic capsule, behind it overlaps the parietal, and above and behind articulates with and partly overlaps the frontal. Anteriorly it is joined to the outer edge of the alisphenoid, and has articulated to its lower angle the quadrate.

n. The *parietals* are a pair of almost square bones, meeting along the middle line of the roof of the skull, and covering in the greater part of the brain cavity. They articulate anteriorly with the temporals, externally with the squamosals, and posteriorly with the supra and exoccipitals. Along the median line they are united to one another. In old specimens they are so firmly anchylosed to surrounding bones that they cannot be detached.

o. The *periotic capsule*, composed of the proötic, epiotic, and opisthotic, is seen on the inside of the cranial cavity, its parts being inseparably fused with each other and with the basisphenoid and occipital segment of the skull.

p. The *basi-sphenoid*, lying in the base of the skull in the median line, is massive in front, where it rises to form the *sella turcica*, a concavity bounded posteriorly by a ridge representing the *posterior clinoid processes*. Posteriorly the inner surface of the bone is hollowed to receive the medulla oblongata ; anteriorly it has, projecting from it, a long rostrum, which partly divides the nasal cavity anteriorly, there being no *vomer* in the pigeon. On the ventral surface, where the rostrum joins the body of the bone, concealed by a pointed process, is the common opening of the *Eustachian tubes*. The basi-sphenoid articulates posteriorly with the basi-occipital ; externally with the periotic capsule, squamosal, and parietal; anteriorly with the alisphenoids. The rostrum movably articulates with the palatines and pterygoids, and meets the ethmoid and inter-orbital septum above.

q. Springing from the anterior elevated portion of the basi-sphenoid on either side of the sella turcica are the *alisphenoids*, which constitute the greater part of the posterior orbital walls. Above they articulate with the parietals, externally with the squamosals.

r. The *supra-occipital* forms the upper boundary of the foramen magnum, and articulates in front with the parietals and periotic capsules, externally with the exoccipitals and the periotic capsules.

s. The *exoccipitals* articulate with the supra-occipital above and with the basi-occipitals below, and are in close relation with the periotic capsules. They form the lateral boundaries of the foramen magnum and enter into the occipital condyle.

t. The *basi-occipital* articulates with the basi-sphenoid in front and completes, posteriorly, the floor of the skull. It forms the inferior boundary of the foramen magnum, enters into the occipital condyle, and unites with the exoccipitals externally. The elements of the occipital segment are very early anchylosed.

22. Make with a fine jeweller's saw a median, longitudinal dorso-ventral section of another skull, and on its inner side note—

a. The *basi-occipital* behind, divided in the median line.

b. External to the basi-occipital, the *exoccipital*, having in it the foramen for the twelfth (*hypoglossal*) nerve (131, *k*).

c. Further out, between the exoccipital and periotic capsule, the *foramen*, through which passes the ninth, tenth, and eleventh nerves (131, *j*).

d. Above this, in what may be called the petrous portion of the periotic capsule, the *foramen* for the eighth nerve (*meatus auditorius internus*) (131, *h*), and anterior to it the foramen for the seventh nerve (131, *g*).

e. Above and anterior to the latter, situated on the ridge which separates the fossa for the medulla from that for the optic lobes, is a cavity which receives the *Gasserian ganglion* (131, *f*).

f. In the basilar portion of the basi-sphenoid a small foramen, which is the commencement of the canal by which the sixth nerve passes through the floor of the skull to reach the orbit.

g. On the side of the skull, behind, may be seen the whole of the *posterior vertical, semi-circular canal* and parts of the *anterior vertical* and *external horizontal canals*.

h. Above the excavation which lodges the optic lobe, note the great fossa in which is placed the *cerebral hemisphere*.

i. Behind the periotic capsule, and covered in by the supra-occipital, the *fossa for the cerebellum* is marked by transverse ridges corresponding to the sulci of the cerebellum.

j. Anteriorly may be seen the *nasal cavity* bounded internally by the perpendicular plate of the ethmoid.

23. The Mandible.—The lower jaw is a V-shaped bone, composed on either side of a *body* and *ramus*, the former compressed laterally until it is reduced to a thin, vertical plate, presenting in its middle a cartilaginous interspace (except in old specimens), which is a portion of Meckel's cartilage. The ramus is expanded laterally, and on its upper surface has a double articular cavity, into which the condyles of the quadrate fit when the mandible is in place. Beneath the articular surface the ramus is continued downward into a process which is also directed outward.

24. Boil the bone carefully for some time, and then, seizing the body between the thumb and forefinger of the left hand, with the thumb and forefinger of the right twist the ramus inward, working it from side to side gently, and it will be found to separate along the upper margin of the cartilaginous interspace and along a line which curves outward toward the outer part of the ramus. Now remove—

a. The outer portion of the proximal segment, which is flattened vertically, enlarged and curved outward at its posterior end, and has its anterior extremity reduced to a thin lamella, which fits between two other thin plates of bone projecting from the dentary (24, *e*). This is the *surangulare*.

b. Beneath the exposed portion of Meckel's cartilage, applied to the inner side of the mandible, is a thin plate of bone, the *splenial*.

c. Between the lower border of the surangulare, on the outside, and the splenial, on the inside, is a small, flattened bone, becoming rounded posteriorly; this forms part of the inferior border of the lower jaw, and is the os *angulare*.

d. *Meckel's cartilage* will now be seen to terminate posteriorly in an ossified mass, which carries the articular surface of the mandible, and constitutes the os *articulare*.

e. The portion of body now remaining is the *dentary*, which unites with the corresponding bone of the other side at the symphysis.

25. The **Hyoid** is composed of a *body* vertically flattened in front and tapering to a point behind; articulated to it, at the junction of its anterior and middle thirds, is a pair of long, slender horns, which lie parallel to the posterior border of the mandible, beyond which they extend to curl upward around the back of the neck. Some distance behind their middle they are marked by a joint. The body of the hyoid is continued forward into the tongue.

26. The **Cervical** Vertebræ are fourteen in number. Take one, say the seventh, for more special study. On it note—

a. The *centrum* and *neural arch*; the former is much elongated antero-posteriorly, flattish ventrally and on the sides, and somewhat constricted about the middle.

b. The anterior end of the centrum is saddle-shaped, being convex dorso-ventrally and concave from side to side. Its posterior end has a similar surface, but concave dorso-ventrally and convex from side to side.

c. From the anterior part of the centrum on each side arises a process perforated at its base. Through the discontinuous tube formed by these processes of successive vertebræ pass the cervical sympathetic nerve (149), and the vertebral vein, and artery (74 ; 88, *a*). Projecting posteriorly and somewhat outward from the process is a slender spicule. The whole process probably represents a rudimentary rib anchylosed to the vertebra by its capitular and tubercular processes.

d. From the inner margin of the process projects mesially and somewhat ventrally a small, bony plate, which, with its fellow, partially covers over a groove on the anterior half of the body of the vertebra in which the carotid arteries (76) normally lie.

e. The *prezygapophyses* spring from the junction of centrum and neural arch. Their articular surfaces are directed dorsally and mesially.

f. The *neural arch* is emarginated in front, and especially behind ; hence between succeeding vertebræ diamond-shaped apertures are left which lead into the neural canal.

g. A pair of long *postzygapophyses* are borne by the neural arch. They are directed posteriorly and outward, and bear articular facets, which look ventrally and outward. Fit a few vertebræ together, and see how their articulating surfaces meet.

27. Most of the cervical vertebræ resemble that just studied; some few, however, require special notice :

   a. The **Atlas** is a small, ring-like bone, presenting, toward its ventral portion, two cup-like articular sockets, an anterior and posterior. In the natural position of the parts of the skeleton the occipital condyle fits into one of these and the odontoid process of the axis (27, *b*) into the other. These sockets are only separated by a thin membrane, and their dorsal boundary is formed by the *transverse ligament* which crosses the ring of the atlas and separates the neural canal from them. A pair of small postzygapophyses spring from the neural arch.

   b. The **Axis**, or *second cervical vertebra*, has a small centrum, much compressed laterally and posteriorly produced ventrally so as to form a large median process. Attached to it in front is the peg-like *odontoid process.* The neural arch is not emarginated posteriorly, and is surmounted by a *spinous process.* At the junction of the centrum and neural arch, in front and behind, is situated a pair of articular processes (*zygapophyses*) ; the anterior are very small. Ventral to them lie two grooves for the passage of vessels going to the brain, which here enter the spinal canal.

   c. The *third, fourth, eleventh, twelfth, thirteenth,* and *fourteenth cervical vertebræ* have distinct spines projecting from their centra in the median ventral line. The thirteenth and fourteenth carry well-developed ribs, which, however, do not unite with the sternum or sternal ribs; the rib of the fourteenth sends backward from the lower third of its posterior border a large process. The twelfth, thirteenth, and fourteenth have well-marked dorsal spinous processes.

28. The **Dorsal Vertebræ** are five in number and are firmly anchylosed together to form a single mass. Their centra are much compressed laterally, and those of the first three prolonged ventrally to form large spines. The neural arches are surmounted by a high ridge composed of the coalesced spinous processes. On each side the centra have articular facets for the heads of the ribs, and springing from the sides of the neural arches are seen the *tubercular processes*, against which the angles of the ribs abut, almost united to form a plate overlying the proximal portion of the ribs on each side.

29. The Lumbar Vertebræ, three in number, present broad, massive centra; they have well-marked transverse processes which are united to the ilia (47) on either side. The *centra* are firmly anchylosed with each other and with the dorsal and sacral vertebræ. They present dorsally a well-developed spine, which is continuous with the similar spine of the last dorsal vertebra.

30. The Sacral Vertebræ, four in number, strongly resemble the lumbar; the fourth is furnished with

THE RIBS.

transverse processes, remarkable for their great length, which unite with the ilia. The first three have only processes corresponding to the tubercular transverse processes of the dorsal vertebræ, and similar to those of the caudal vertebræ.

31. The **Caudal Vertebræ**, twelve in number, may be divided into *six fixed* and *six movable*.

    a. The fixed, anchylosed to the sacrum and to one another, have horizontal, broad, transverse processes united to the ilia and to each other, which may be readily distinguished as the portions of bone between them are thin and translucent. Running along the dorsal surface of their neural arches is a rounded elevation, which is a continuation of that in the dorsal and sacral region, and formed by the spinous processes of the coalesced vertebræ.

    b. Following the fixed are five movable caudal vertebræ, having large spinous and transverse processes.

    c. Articulated to the end of the caudal region is a ploughshare-shaped bone (*pygostyle*), formed by the coalescence of a number of vertebræ, which supports the uropygium.

32. The **Ribs** are seven on each side; two belong to the cervical region and have been described (27, *c*). The remainder belong to the dorsal region and are each composed of two parts, a dorsal and a ventral, called respectively *vertebral* and *sternal ribs*.

    a. The vertebral rib is a broad lamella, ossified throughout, and having at its dorsal end a process (*tuberculum*) articulating with the transverse process of the corresponding vertebra. On the ventral surface of its dorsal end is a vertical ridge, becoming free and rounded opposite the tuberculum, and passing downward and inward to form the *capitulum* which articulates with the vertebral centrum.

   b. From the posterior margins of the four anterior pairs of vertebral ribs, large processes project upward and backward (*uncinate processes*).

   c. The *sternal ribs*, with which the vertebral ribs articulate at their outer ends, are also thin and flattened, the four first being articulated to the outer margin of the sternum (33, *b*), near its anterior end. The fifth is shorter than the fourth, and at its sternal end anchylosed to it, having no direct connection with the sternum.

33. The **Sternum** is remarkable for its great size.

   a. Anteriorly it presents two long, deeply grooved, articular surfaces for the inner ends of the coracoids (35), and between and dorsal to them a process (*rostrum*) which is bifurcated.

   b. The lateral borders are for some distance almost parallel, but converge somewhat posteriorly. On them are seen, just behind their anterior angles, the articular surfaces for the sternal ribs (32, *c*), followed by a large process (*middle xiphoid process*) pointing backward and outward. Behind this is a deep notch (*fontanelle*) filled by membrane in the recent state and succeeded by a second more slender process (*external xiphoid process*) which fuses at its distal end with a thin bar of bone, forming part of the posterior border of the sternum ; the two surround a second smaller, triangular fontanelle.

c. The posterior borders of the bone form a more obtuse angle than the anterior and are largely formed by the bony bars on each side just mentioned.

d. Projecting ventrally in the median line is a great bony *keel* (*lophosteon*), which, together with the ventral surface of the sternum proper, affords an origin for the powerful muscles of flight (64).*

34. The **Shoulder Girdle** is made up of a *coracoid*, *clavicle*, and *scapula* on each side.

35. The *coracoid* is a short bone, having a somewhat rounded and flattened shaft expanding into a broad, proximal extremity, which presents a ridge for articulation with the groove upon the anterior border of the sternum (33, *a*), and sends backward and outward a large process beyond it.

36. The distal or glenoidal end of the coracoid is irregular in form. On its anterior surface is a deep groove leading to a foramen for the tendon of the pec-

---

* The sternum of the Carinatæ is developed from five centres of ossification, giving rise to five distinct and separable bones in the young pigeon, viz.: the *keel*, called lophosteon, an antero-lateral pair, pleurostea, and a postero-lateral bifurcated pair, metostea, separated from the pleurostea by a line passing inward just behind the articular surfaces of the ribs.

toralis minor (67), which passes through the bone. Above this foramen is a large process with which the clavicle is articulated; below the foramen, on the inner dorsal aspect of the bone, is a rough surface to which the scapula is attached; to its outer side is seen the dorsal half of the *glenoid fossa*.

37. The *scapula* anteriorly is expanded laterally and presents an articular surface which forms the lower half of the glenoid cavity; internal to this is the surface by which it meets the coracoid, and which is continued forward to form a process, which in the recent state is united by ligament to the coracoid behind the attachment of the clavicle. Posteriorly the scapula is flattened dorso-ventrally and lies over the ribs.

38. The *clavicle* is a thin bar of bone attached dorsally to the coracoid, and extending thence ventrally and mesially to meet its fellow on the middle line, where the two fuse to form the *furculum*. At their junction they give origin to a process (*hypocleidium*) flattened from side to side, and normally united by ligament to the keel of the sternum.

39. The **Bones of the Fore Limb,** when in a position of rest, lie in a nearly vertical plane, and are folded in such a manner that the humerus is parallel to the long axis of the body; the ulna and radius flexed upon the humerus in a position midway between pronation and supination, and the manus upon the bones of the forearm, so that the thumb is directed ventrally.

40. The *humerus* is a stout bone articulating anteriorly with the glenoid cavity by a vertically elongated

articular surface, ventral and to the inner side of which is a large process having its posterior aspect hollowed out to form the *pneumatic cavity*, which receives a prolongation of an air sac (85). On the dorsal and outer surface of the head of the bone is a large triangular process (*greater turbosity*). The shaft is short, thick, and somewhat flattened; it widens out behind, where it passes into the *external* and *internal condyles* which bear the articular surfaces for the ulna and radius.

41. The *radius* is a long, straight, cylindrical bone, expanded proximally to form a small head which bears a cup-shaped cavity for articulation with the humerus: by its outer posterior margin, it articulates also with the ulna. Posteriorly, the shaft terminates in a slightly expanded head which articulates with the ulna and radial carpal bones.

42. The *ulna* is a larger and thicker bone than the radius and considerably bowed outward. Its outer border exhibits a row of small prominences, which mark the points of attachment of the secondary remiges (9, *d*). Its proximal end articulates with the radius and humerus, and its distal end with the radius and the radial and ulnar carpal bones, and metacarpal bone of the index digit.

43. The *carpal bones* are two squarish bones, placed on either side of the head of the metacarpal bone of the index digit when the hand is extended, but in the flexed state one lies ventral to the distal end of the ulna, the other between it and the radius. They articulate proximally with the radius and ulna, and distally with the three metacarpal bones.

44. The *metacarpal bones* are, proximally, firmly anchylosed together, and differ much in size and form; the first, that of the pollex, is very small and short and has a large process directed forward; the second is straight, thick, and much stouter and longer than its fellows; the third is slender, flattened, and shorter than the second, to which it is anchylosed at its distal extremity also, the interval between the shafts of the second and third metacarpal bones being in the recent state occupied by membrane.

45. The *digits* are also peculiar. The pollex has one phalanx, which is short and pointed at its distal extremity; the second digit has two phalanges, and is much larger than the thumb; the first is expanded into a broad bone, which thins away to a sharp edge on its posterior margin; the third digit has but one phalanx, irregularly triangular in form.

46. The **Pelvic Girdle** is composed of the ilium, ischium, and os pubis on each side.

47. The *ilium* is greatly elongated antero-posteriorly and firmly united to the transverse processes of all the lumbar (29), sacral (30), and anchylosed caudal vertebræ (31). Anteriorly it overlaps the last rib, with which it is also anchylosed. It is divided into two parts—an anterior dorsally concave and posterior dorsally convex portion; the former contributes at its posterior end to form more than half of the *acetabulum*, just above which it presents a plane articular surface (*antitrochanter*), looking ventrally and forward, against which the great trochanter of the femur (51) plays.

48. The *ischium* is nearly vertically placed beneath the hinder portion of the ilium. In front it forms about one-fourth of the acetabulum, and, widening as it extends backward, is soon united to the ilium dorsally, and pubis ventrally, with the former surrounding the *ilio-sciatic foramen*, and with the latter a smaller foramen for the passage of the tendon of the obturator internus muscle. The bone behind the ilium slopes away ventrally and backward for some distance. A narrow slit exists between the middle portion of the ischium and the pubis ; this is normally closed by membrane and constitutes the *obturator foramen*.

49. The *pubis* is a narrow bar of bone forming the lower fourth of the acetabulum, and arching outward is united to the ischium in front of and behind the obturator foramen. Its distal end is widely separated from its fellow of the other side.

50. Note that the inner wall of the acetabulum is wanting, the opening thus left being filled by membrane during life.

51. The **Bones of the Hind Limb.** The *femur* has a rounded shaft, slightly bowed forward. At its upper extremity it is laterally expanded and on its inner side bears a globular head, placed at right angles to its shaft. Opposite the head it is extended to form the *great trochanter*. Its distal extremity widens out to form an *inner* and *outer condyle*, which project backward considerably beyond the line of the shaft and have between them a deep groove extending from the front to the back of the bone. Upon the posterior inferior surface of the outer condyle is a sharp ridge which plays between the heads of the fibula and tibia.

52. In front of the knee-joint a sesamoid bone, the *patella*, is developed in the tendon of the extensor muscles. It will probably have been lost in cleaning the skeleton.

53. The *fibula*, a slender bone, only imperfectly developed, does not extend below the middle of the tibia, along the outer side of which it is placed. Above it is expanded into a head, which, by its upper surface, articulates with the outer condyle of the femur, and by its inner surface with the head of the tibia. About its middle for some distance it is anchylosed to the shaft of the tibia.

54. The *tibio-tarsus* is remarkably long, and consists not only of the tibia, but of the proximal bone of the tarsus, which becomes fused with it at an early period, being represented by a separate cartilage in the embryo. Its shaft is long and cylindrical. Above it expands to form a broad surface with which the greater part of the outer and all the inner condyles of the femur articulate. Projecting from the front of the head is the *cnemial process*. The lower extremity is terminated by an articular surface for the next bone.

55. The *tarso-metatarsus*, which is formed by the fusion of the *second, third*, and *fourth metatarsal bones* with each

other, and with the *distal division of the tarsus*, which, in the embryo, is represented by a separate cartilage. The metatarsal bones do not lie side by side, the third having its head pushed backward, and its lower extremity somewhat in advance of its fellows. The proximal end of the third metatarsal is prolonged backward to form a large process, on the inner side of which is a deep fossa. The distal extremities of the metatarsals remain distinct and present double articular surfaces for the proximal phalanges of the digits. The first metatarsal bone, that of the hallux, is small and united by ligament to the inner posterior aspect of the second metatarsal near its distal end.

56. The *digits of the pes*, four in number, are each terminated by a stout, curved claw. The first (*hallux*) has two phalanges; the second, three; the third, four, and the fourth, five. The third digit is much longer than the fourth, and the second and fourth of about equal length. The hallux is directed backward and inward, the second, third, and fourth forward.

57. General Dissection.

Carefully divide the skin along the median line from the lower mandible to the cloacal opening, and reflect it as far back as the anterior extremity of the sternum. Before doing this, it is well to pass a glass tube into the throat of the bird, and blow air down the œsophagus, so as to inflate the crop and get a correct idea of its situation, as otherwise its delicate walls may be injured in removing the skin. The air may be prevented from escaping by a string tied around the anterior portion of the neck.

53. Passing down the *left* side of the neck will be seen the *trachea*, which may be traced till it disappears in the pleuro-peritoneal cavity; and on the *right* side the *œsophagus*, which is extremely distensible and thin, its walls being folded longitudinally. At first it is dorsal to the trachea, but passes very soon to the right, and runs backward to end in the middle compartment of the trilobed *crop*, at its entrance into which it is somewhat narrowed.

59. If the bird had been rearing a brood the walls of the lateral pouches of the crop will be seen to be thickened, owing to the activity of *glands*, which secrete a fluid to act upon the food and soften it. Carefully tease away the crop from its attachment to the anterior border of the *pectoralis major* and furculum, taking great care not to injure the *air sacs* which lie behind it.

60. Allow the air to escape from the œsophagus and crop, and inserting the small end of a pipette into the larynx, inflate again, when the *air sacs* will be seen to fill. Above the anterior end of the sternum, and lying along the limb of the furculum on each side, will be seen *a sac*, and above and between these, ventral to the trachea and œsophagus, *a third*.

61. Beginning behind the mandible, open the œsophagus as far as its entrance into the crop, and also the crop for a short distance. If the organ contains grain clean it out, so that its interior may be examined, and the point where the œsophagus begins again beyond it.

62. Observe, attached to the ventral side of the

trachea and passing outward, backward, and downward, a pair of delicate muscles whose origin will be seen later (70).

63. Occasionally moistening the neck to prevent drying, proceed to reflect the skin from one side of the breast and from one fore limb while it is still fresh; hold it up to the light and examine the muscles which surround the feather follicles and serve to erect the feathers. The *great pectoral muscle*, which forms the chief bulk of the fleshy breast, will thus be exposed in part.

64. Make an incision in the left *pectoralis major muscle* about an eighth of an inch from the keel of the sternum, deepen the cut until the muscle is divided and the muscles beneath reached, which will be known by the glistening appearance of the fascia. Reflect the great pectoral and cut through its origin from the furculum and from the membrane uniting the furculum to the sternum; near the shoulder-joint will be seen an *air sac*, one of those already noted at the base of the neck (60). Next, cut away the great pectoral where it arises from the ribs and outer posterior part of the sternum, noting the distinct bundles which it presents; and, raising it, follow it to its insertion on the greater tuberosity of the humerus, taking care to avoid severing the *artery, vein, and nerves* seen entering it on its upper surface.

65. Follow the nerves until they are seen to be given off from the *brachial plexus* and trace the vessels until they disappear in the pleuro-peritoneal cavity.

66. From its attachment follow back the tendinous septum (*aponeurosis*) of the great pectoral muscle; it will be seen that it divides the muscle into two portions: an *outer*, arising from the outer part of the ventral surface of the sternum, and an *inner*, arising from the keel of the sternum, the furculum, and the membrane joining the furculum to the sternum and coracoid. Note that the outer is turned under the inner division of the muscle.

67. The *pectoralis minor* is now exposed in almost its whole extent. It arises from the sternum between the two divisions of the pectoralis major and from the membrane which unites the furculum to the sternum and coracoid. Raise it and follow it as it runs along the inner margin of the coracoid until it terminates in a tendon, which passes through a foramen situated partly in the distal end of the coracoid and partly between the scapula and coracoid at their point of meeting; the tendon then running backward and outward, is attached to the ridge on the upper and inner aspect of the greater tuberosity of the humerus. Carefully clear away the muscles about the shoulder-joint and dissect out this insertion. After the wing has been depressed by the great pectoral the pectoralis minor elevates it; this can be proved by replacing the muscles in their natural position and pulling on them.

68. An opening will now be exposed, bounded by the sternum ribs, scapula, and coracoid, in which may be seen the *innominate artery*, giving off, first, the *carotid*, and then becoming the *subclavian*, which latter immediately divides opposite the outer angle of the distal ex-

tremity of the coracoid into the *brachial* (which disappears among the muscles of the arm just posterior to the shoulder-joint), and the *pectoral*, which splits into two branches and supplies the pectoral muscles. Accompanying these, and lying beneath them, may be seen the corresponding *veins* bringing back blood from the same parts and uniting to form the *left innominate vein*.

69. Descending from the region of the neck, and running outward and backward, the nerves which form the *brachial plexus* may be seen sending branches to the surrounding muscles and terminating finally in two trunks, the anterior of which (the *circumflex*) curls around the shaft of the humerus, the other, the *brachial*, passes down the posterior aspect of the arm.

70. If the dissection has been carefully made, a small, delicate muscle will be exposed within the thoracico-abdominal cavity, which, arising from the dorsal surface of the sternum close to the articulation of the first rib and passing forward, is attached to the trachea, as already seen (62).

71. **Dissection of the Neck.**—Note on the side of the neck the long thyroid body extending from behind the articulation of the mandible with the skull to the root of the neck.

72. Ventral to this, the *jugular vein* and *pneumogastric nerve*, both of which should be traced upward to the base of the skull (74 ; 145), and down the neck until seen to enter the thoracico-abdominal cavity. If the vein is

not apparent allow the neck to hang over the edge of the table and it will soon become distended with blood.

73. Internal to the articulation of the jaw will be seen the ninth nerve (146) giving off a branch (*descendens noni*), which runs down the neck by the side of the jugular vein.

74. Divide the œsophagus and trachea about an inch posterior to the head, and turn them forward, carefully dissecting them away from the parts which lie above them, then proceed to trace the jugular vein; near the base of the skull will be found an arch formed by the union of the jugular veins of both sides into which several vessels enter, bringing blood from the regions supplied by the carotid arteries. At the base of the neck they are again united by a large cross-branch, and receive on either side the vertebral veins and the veins of the ingluvies.

75. Note the *cervical spinal nerves*, which supply the various structures of the neck. There are fifteen pairs in all. They pass out from the spinal canal in front of the bodies of the cervical vertebræ through the foramina beneath their articular processes, with the exception of the last (fourteenth), which issues posterior to the thirteenth vertebra of this region.

76. Divide carefully in the middle line the muscles on the ventral side of the cervical vertebræ; the *carotid arteries* will then be seen running close together on the ventral sides of the vertebral centra, the downward

processes of which (26, *d*) form a groove in which the arteries lie. Opposite the fourth cervical vertebra they diverge, each passing to its own side, and dividing behind and a little to the inside of the articulation of the lower mandible into—

    a. The *external carotid*, which supplies the outside of the head and floor of the mouth.

    b. The *internal carotid*, which enters the skull through a foramen in its base, anterior to the foramen of exit of the seventh nerve, and supplies the brain.

77. Clear away the muscles on the dorsal surface of the neck and expose the ligamentum nuchæ which, passing from the anterior border of the neural arch of one cervical vertebra to the posterior border of the vertebra anterior to it, reaches from the fourteenth to the second vertebra of this region. It is composed of yellow elastic tissue, and serves to curve the neck backward and support, in part, the weight of the head.

78. Examine the **intervertebral disks** of several of the movable presacral vertebræ, preferably those of the cervical region, and, if possible, make out that they are attached to the vertebræ only at their margins. Toward the centre they thin away and are perforated in the middle, a ligament in most birds passing through the opening to unite the adjoining vertebræ; this, however, is not present in the pigeon.

79. Make an incision in the skin along the inner aspect of the hind limb, and note, issuing from the **sciatic** foramen and passing beneath the adductor

muscles the *sciatic nerve*, which, in the lower part of the thigh, divides into two branches, the *anterior* and *posterior tibial nerves*, which supply the muscles of the leg. Accompanying the sciatic nerve is the *femoral artery* and *vein*, which likewise divide into anterior and posterior tibial branches.

80. Arising by a flat tendon from the lower margin of the acetabulum and running down the inside of the thigh along the anterior edge of the vastus internus is an exceedingly delicate muscle terminating in a long, slender tendon which passes over and is embedded in the patella, then over the outside of the knee-joint, and finally is attached to the aponeurosis of the superficial flexor of the digits of the foot in the calf region. This is the **perching muscle**, and is so arranged that when the bird flexes the leg upon the thigh, as it does involuntarily when falling asleep, the flexor muscles of the toes are pulled upon and the foot made to grasp the perch.

81. **Dissection of Pleuro-peritoneal Cavity.**—Carefully cut away the abdominal muscles from the margin of the sternum, except in the middle line behind, where a band must be left a quarter of an inch wide, uniting the sternum and pubes. From these longitudinal cuts next make on each side two incisions, one in front and one behind, and each reaching dorsally to near the vertebral column. The flaps so made may be turned back, and a view will be had of the inside of the abdominal cavity without injuring any of the air sacs.

82. Holding the keel of the sternum in one hand,

AIR SACS. 133

with the other insert a glass tube into the cut end of the trachea and inflate. Behind, in the median line, an air sac will be seen to fill the **abdominal air sac**, and in front and above this, on either side, two **thoracic sacs** can be made out by first repeatedly inflating them and allowing them to collapse, and then carefully making windows in their walls and exploring their interior.

83. Lying above the thoracic air sacs can be seen *the lungs*, with which they communicate at their inner border. The abdominal air sac communicates with the lungs at the posterior extremities of the latter. The anterior thoracic sac sends a prolongation forward and upward toward the cervical region.

84. Clip the walls of the air sacs, and cut through the articulations of the vertebral and sternal ribs on both sides, taking great care to preserve the fold of peritoneum, which, running from the sternum and dorsal surface of the heart, passes upward and backward to meet the abdominal walls, and upward between the lobes of the liver to be lost among the viscera; in it will be seen a vein which, in front, passes into the liver between its two lobes, and behind dips into the pelvis.

85. Raise the sternum, and at the same time carefully detach the heart from it, and a **single unpaired air sac** will be seen just beneath the anterior portion of the sternum; this sac is partially divided by a partition in front, and when seen from the outside appears double. A diverticulum from it extends to the pneumatic opening on the under surface of the head of the

humerus, and it communicates with the air cavities of the sternum by means of the opening on the superior surface of the latter in the middle line near its anterior end. External to this air sac on either side is an **air sac** already seen on the inner side of the shoulder-joint and limbs of the furculum (60).

86. Now cut the ligament and membrane, binding the furculum to the sternum; disarticulate the inner ends of the coracoids, severing carefully the ligaments on their upper surface attaching them to the sternum, turn the coracoids forward and outward, and study the relative position of the organs occupying the abdominal cavity.

   a. In front, in the median line, the large, muscular, cone-shaped **heart** (the apex of the cone being directed backward), enclosed in a thin, transparent membrane (*pericardium*) which extends for some distance upon the vessels arising from its base. Prick this, and with a pipette inflate it. The right and left *auricles* can be easily distinguished from the *ventricles* by their dark color, and are marked off by lines of fat from the ventricles, and from each other. The line of division between the ventricles is not evident.

   b. Behind the heart, the **large liver** with its right and left lobes, each extending forward some distance and partly concealing the heart. The *right lobe* is much the larger of the two and extends dorsally far into the abdominal cavity, and backward almost as far as the posterior extremity of the sternum, and across the middle line. The *left lobe* is comparatively small and extends upward on the left side.

c. On the left side, behind the left liver lobe and partly covered by it, the **gizzard** (*gigerium*) presenting a rounded, pearly-looking surface, when viewed from the left side.

d. Turn forward the liver, and dorsal to it will be seen the coils of *intestine* covered by peritoneum, and having a general oblique direction from right to left, and from before backward.

e. Between two coils of intestine, just behind the gizzard, a portion of the *pancreas* appears.

87. Now proceed to study the vessels connected with the heart. Clean away the pericardium, and trace some to the left ventricle, where they will be found to arise from a single trunk, the **aorta**, which, after leaving the heart, almost immediately gives off the **right and left innominate arteries,** and then turning to the right and passing over the right bronchus continues as the *dorsal aorta.*

88. The innominate on either side gives off—

a. The **carotid**, which, running forward to the ventral surface of the neck, soon meets its fellow in the median line and pursues the course already described (76). Opposite the shoulder-joint it gives off the *vertebral artery*, which enters the vertebral canal behind the tenth cervical vertebra, and immediately afterward the *ascending cervical,* which runs forward along the side of the neck close to the pneumogastric nerve. In the angle between the latter branch and the main trunk may be seen a small elongated red body, the **thymus gland**.

 b. The **subclavian**, which soon becomes the *brachial*, and crosses the axilla to reach the wing.

 c. The residue of the innominate then becomes the **pectoral** and almost at once divides into two branches, which supply the pectoral muscles (68).

89. Raise the apex of the heart and turn it forward, remove the peritoneum and pericardium, which conceal the veins entering it, exercising the greatest care not to tear them, for, if this should happen, it will not be easy to make out all the following points:

 a. Crossing the dorsal surface of the heart at the junction of the left auricle with the base of the left ventricle is the large **left** superior cava, which is formed by the junction of the jugular and subclavian veins, and empties into the right auricle at its inner side. The vein, as it lies on the heart, receives two branches from the proventriculus. Inflate these vessels through the cut extremities of the pectoral veins or any other convenient branch.

 b. Issuing from the right lobe of the liver, where it receives the hepatic veins, will be seen the **posterior cava**. Make a small opening in this and inflate the heart through it, when an excellent view of all its parts can be obtained. The inferior cava enters the right auricle on its dorsal surface.

 c. Entering the right auricle on the right side

near its anterior surface will now be seen the **right superior cava**, receiving the blood from vessels similar to those which make up the left.

90. Before making any further dissection, examine the **anterior abdominal vein**, which is formed by the union of two veins in the median line behind, which lie on the ventral surface of the muscles of the uropygium. It then runs ventrally and forward, and unites with the portal vein in the neighborhood of the pyloric portion of the stomach.

91. The **portal vein** should be traced from its commencement between the folds of the mesentery, near the gizzard, until it is found to enter the right lobe of the liver.

92. The **Posterior Cava.**—The veins which unite to form the anterior abdominal meet on each side a vein returning blood from the uropygial region, which passes ventral to the kidneys, receiving—

   a. A branch from the posterior lobe of the kidney.
   b. The *femoral vein*, which passing between the middle and anterior lobes receives the blood from the latter and from the hind limb.
   c. A branch from the anterior lobe of the kidney.

It then unites with its fellow to form the posterior cava, which passing to the right of the proventriculus enters the right lobe of the liver, and after receiving the blood from both right and left lobes empties into the right auricle, as already seen.

93. Cut the innominate arteries and turn them forward; above them will be seen the dorsal **aorta**, first

passing to the right and then curling over the right bronchus to get to the right side of the vertebral column.

94. On the right side of the œsophagus is the **right pneumogastric nerve**, which must be cleaned; just in front of the aortic arch it will be seen to give off the *recurrent laryngeal branch*, which, passing behind the arch of the aorta, runs forward along the right side of the œsophagus, to which it gives branches. Branches also run to the right lung and to the heart from the main trunk, which then runs backward to the stomach, and must be followed later (98).

95. Cut the aortic arch and follow the **pulmonary artery**, which arises from the base of the right ventricle, and immediately divides into right and left branches, which passing ventral to the bronchi enter the lungs about the middle of their ventral surface, external to the bronchi.

96. On the left side of the trachea will be seen, by clearing away the connective tissue of the part, the **left** pneumogastric nerve, which just in front of the left pulmonary artery, near the point where it enters the lung, gives off its *recurrent laryngeal branch*, which turning round the innominate artery runs up the neck again to the inner side of the main trunk, and is distributed in the same manner as the right recurrent laryngeal. The left pneumogastric gives branches to the lung and to the heart, and passing to the proventriculus runs along its under surface to the gizzard (106), upon the left side of which it is distributed.

97. Divide the innominate and pulmonary arteries on each side. Raise the anterior end of the heart and find the **pulmonary veins**, which leaving the lungs internal to the entrance of the bronchi enter the left auricle on its dorsal side.

98. Cut the right pulmonary vein, turn the heart over to the left, and note on the ventral surface of the proventriculus the stomach branch of the *right pneumogastric*, which may now be traced from the point where the right recurrent branch was given off. It will be found to pass over the right bronchus dorsal to the vessels proceeding from and entering the heart, and passing to the right side of the proventriculus to be distributed to the right side of the gizzard.

99. Study now the **dorsal aorta.** It will be found that it curls around the right bronchus, and passing along the right side of the œsophagus reaches the right side of the vertebral column, and then soon passes to the median line. Its branches are:

    a. The bronchial arteries.
    b. Several *intercostal* branches on either side.
    c. The **cœliac axis,** which passing along the right side of the proventriculus gives off—

        a. A large, short trunk, which reaching the ventral side of that organ divides into two branches. One passes forward along its ventral surface and the other supplies the left side of the gizzard, along which it may be traced.

        β. Several small twigs to the *spleen*, which is an elongated dark red body closely attached to the vessel.

γ. A branch to the right side of the gizzard.

δ. A small branch to the liver, which enters it with the portal vein.

ε. The vessel finally terminates in a branch which supplies the pancreas lying between the parallel coils of the duodenum.

d. The **superior mesenteric**, which passes to the intestines between the folds of the mesentery, there giving off branches to form arches, from which smaller twigs arise to pass to the gut.

e. The *iliac*, which runs outward between the anterior and middle lobes of the kidney, and passes over the brim of the pelvis just above the acetabulum to enter the hind limb on its inner anterior aspect, where it becomes the femoral and runs beneath the sartorius muscle.

f. The **sciatic**, a large branch on either side which passes outward and backward between the middle and posterior lobes of the kidneys, to both of which it gives branches, and issues from the pelvis through the sacro-sciatic foramen. It then pursues its course down the inner posterior aspect of the hind limb, of which it is the main artery.

g. The *dorsal aorta* finally terminates as the *inferior mesenteric*, which runs to the rectum and anastomoses with the branches of the superior mesenteric.

100. Note the network of ganglia and nerves (*solar plexus*) near the origin of the cœliac axis and superior mesenteric arteries and the nerves which arise **from it**

and accompany the arteries to be distributed to the abdominal organs.

101. The Vocal Organs.—Note at the posterior extremity of the trachea and the beginning of the bronchi, into which it bifurcates posteriorly, the **syrinx** or *second larynx*, in which the voice of the bird is formed. Its construction is as follows :

   a. The three last tracheal rings are united by cartilage along the median ventral and dorsal lines ; the last two rings are widely separated ; these form the *tympanum of the syrinx*.

   b. The bronchial cartilages are wanting on the inner side of both bronchi, their place being occupied by a membrane (*membrana tympaniformis interna*) which anteriorly rises up between the openings of the bronchi into a free fold (*membrana semilunaris*).

   c. Arising from the ventro-lateral aspect of the trachea, two delicate muscles run backward and outward to be attached to the sides of the tympanum.

   d. A second pair of muscles arise from the ventral aspect of the trachea between the pair just described and run backward and downward to be attached to the anterior margin of the sternum close to the articulation of the first rib (62 ; 70).

102. Cut the trachea across just in front of the attachment of the syringeal muscles, slit the tympanum along one side and also one of the bronchi, and note through the opening the free fold (*membrana semilu-*

naris), which is considerably thickened along its free border and which can be best seen by drawing the trachea forward.

103. Cut through the bronchi near the lung and remove the syrinx. Examine the **trachea,** noting that its cartilaginous rings are continuous around its entire circumference but much thinned on the dorsal side.

104. The dissection of the **Larynx** may be made now, or postponed until the completion of the dissection of the cranial nerves, which will otherwise require a fresh bird. Remove about an inch of the trachea and œsophagus, with the hyoid bone, tongue, and larynx, and note—

   a. The attachment of the œsophagus in front to the horn of the hyoid and sides and posterior margin of the arytenoid cartilages. Behind, above, and on the sides it becomes continuous with the mucous membrane of the buccal cavity.

   b. The position of the larynx above the body of the hyoid (25).

   c. Note the shape of the entry to the larynx (*aditus laryngis*) when its sides are separated, square in front and tapering away to a point behind.

   d. Bounding this on either side are the *arytenoid cartilages,* joined behind in the median line but separated in front, where their free ends are united by fibrous tissue to the tip of the thyroid cartilage.

   e. On either side may be seen the muscles, which arising from the upper edge of the thyroid cartilage pass outward and forward to be attached to the outer side of the arytenoid cartilages. When they contract they separate the arytenoid cartilages anteriorly and open the aditus of the larynx.

f. Slit open the trachea and larynx along the median dorsal line, and spreading out the halves of the larynx note the thin *thyroid cartilage*, which, narrow above, below extends forward into a point. The arytenoid cartilages are united to it by a thin membrane.

105. Note the position of the **œsophagus**, which is seen where it lies dorsal to all the organs in the anterior portion of the thoracico-abdominal cavity. Entering the body cavity from the right side of the neck it soon passes to the median line and then somewhat to the left, until it enters the gizzard on the anterior portion of the dorsal border of the latter. As it approaches the stomach it enlarges, its last inch, called **proventriculus**, having thick vascular walls, readily distinguishable from the rest. Near the gizzard it is again somewhat constricted.

106. The **Gizzard** (*gigerium*, or *stomach*) is a hard muscular organ, situated on the left side of the abdominal cavity near its posterior end, and may be compared to a Lima bean, much thickened from side to side, placed on edge with its convex border looking ventrally.

107. Raise the gizzard, dividing at the same time its attachment to the abdominal air sac, which lies dorsal to it; spread out and examine the fold of the *peritoneum* which extends backward from its posterior margin, and note the fat which it contains; this is the *great omentum*, formed by the continuation of the two layers of peritoneum investing the sides of the gizzard. Cut through this and turn the stomach forward.

108. Raise the intestines and observe the delicate *mesentery* which slings them, having between its folds the branches of the mesenteric artery, portal vein, and the lymphatics.

109. On the right side of the proventriculus near its opening into the stomach, note a small elongated dark red body, the **spleen**.

110. The **small intestine** leaving the gizzard on its inner side near the entrance of the proventriculus forms a long loop (*duodenum*) which has in it the large **pancreas**, part of which has already been seen.

111. Trace the **gall ducts** from either lobe of the liver until they enter the duodenum, one near the pylorus and the other the opposite limb of the great loop which contains the pancreas. Observe that *there is no gall bladder*.

112. Unravel the small intestine, cutting it away from the mesentery, and note that it becomes smaller and smaller until it enters the **colon** near the cloaca, the point of junction being marked by two small *cæco-colic diverticula* directed outward and forward. The whole intestine is about nine times the length of the trunk of the pigeon.

113. Remove the whole alimentary canal and the liver, noting the form of the latter and the bridge of hepatic tissue above the apex of the heart connecting its two lobes. The dorsal surface of the right lobe is moulded to fit the depressions between the ribs.

114. Open the *œsophagus* and *proventriculus*. The longitudinal folds and muscular bands of the former are continued to the proventriculus, the interior of which is highly glandular and presents a tufted, spongy appearance.

115. Open the stomach along its inferior border and it will be found to be filled with gravel mixed with grain, which is changed to bright grass-green. Wash away the contents and it will be seen to be lined by a thick, tough membrane presenting numerous folds.

116. Detach the lining membrane and observe that the powerful muscular walls are composed of fibres running in different directions, and that the centre of each wall and the entire inner surface of the organ beneath its lining membrane is composed of dense fibrous tissue which gives origin to the muscular fibres which compose the mass of its anterior and posterior borders. On either side of the cardiac opening and on the border opposite it there are small lateral dilatations. On the right side, just above the *cardiac opening*, is the *pylorus*, which opens into the duodenum.

117. Slit open the small intestine. Observe the longitudinal direction of its folds, which will be found to continue as far as the colon. Find the *duct of the pancreas*, through which pass a guarded bristle into the duodenum.

7

118. Arising from the inner surface of the vertebral ribs near their articulation with the sternal rib, and from the inside of the membrane closing in the space between coracoid scapula and sternum (68), and attached to the bodies of the dorsal vertebræ, is a thin sheet of muscle which is closely adapted to the concave ventral surface of the lungs. This represents the **diaphragm** of mammals.

119. Carefully remove one **lung**, noting the manner in which it adapts itself to the bones on its dorsal side and fills up the spaces between the ribs. Pass a probe into its bronchus, which will be found to pass through the lung to its posterior border, where it communicates with the abdominal air sac. With a scalpel lay open one lung, following the probe, and note the numerous openings in the wall of the main bronchus. Pass a fine probe into several of the larger openings, and they will be found to lead toward the surface of the lung, where some of them communicate with the air sacs. Pass a fine bristle into the smaller openings and they will also be found to pass to the surface of the organ.

120. The only organs now remaining in the abdominal cavity are the reproductive apparatus and the **kidneys**. If it is a cock-bird, note—

    a. Two large, rounded, and elongated yellow organs, **testes**, attached by a fold of the peritoneum to the ventral face of the anterior lobes of the kidneys, and to the centra of the vertebræ adjoining.

    b. The very slender coiled *vas deferens* leaving

the upper surface of each testis near its anterior portion, and running backward over the ventral surface of the kidney beneath the peritoneum to enter the cloaca.

121. Remove one testicle; split it lengthwise, observing that it is invested by a fibrous capsule, and that it presents a uniform white structure, from which a whitish fluid may be pressed. Place this under the microscope, and in it will be seen *spermatozoa* in great numbers.

122. If the bird is **a hen,** note—
- a. On the left side, occupying the same position as the left testicle in the male, the single ovary containing *eggs* in different stages of development.
- b. The *oviduct*, a delicate membranous tube having a funnel-shaped dilatation opening into the abdominal cavity behind and to the outer side of the ovary, from which point it may be traced backward to the cloaca.

123. Lying alongside of the vertebral column will now be seen the dark red flattened kidneys, unequally divided into three lobes, the posterior being the largest and the middle the smallest.

124. Running along the surface of the two hindermost lobes, a bright orange-colored mass, the **adrenal body.**

125. By the side of the vas deferens, and entering the cloaca, is seen a small tube, the ureter; open this, and pass a guarded bristle along it until it enters the cloaca, and in the opposite direction until it is found to have its commencement in the kidney.

126. Divide the **cloaca** along the ventral median line; pass a probe into it through the cut end of the rectum; find the openings of the ureters and vasa deferentia, or Fallopian tubes, as the case may be, by passing bristles through small openings made in their walls where they are seen in the abdominal cavity and pushing them on into the cloaca, which they will be found to enter external to the opening of the rectum. Note the large glandular body attached to the anterior superior wall of the cloaca (*bursa Fabricii*).

127. Dissection of the Heart.—With a pair of sharp scissors divide the remaining vessels entering and leaving the heart, leaving as much as possible of them attached to that organ. Clean away the connective tissue which surrounds them; also the fat which fills up the sulcus between the auricles and ventricles. Trace the aorta until it is seen to arise from the left ventricle. Pass one blade of a pair of delicate straight probe-pointed scissors through the open end of the aorta into the left ventricle, which lies on the left side of the heart, and lay it open along its ventral aspect, keeping the scissors as far to the right as possible, then note the following:

    a.   Just where the aorta joins the ventricle there are three pouch-like cavities (*sinuses of Valsalva*).

    b.   To the edges of these are attached the *semilunar valves*—delicate folds of the membrane lining the heart.

THE HEART.

c. Note in each of these a thickened spot (*corpus Arantii*).

d. On that portion of the wall of the aorta which is immediately above or within two of the sinuses of Valsalva are the openings of the *coronary arteries*, which should be traced along the dividing line between the ventricles; a bristle being passed into each as a guide.

e. At the left of the aortic opening the left *auriculo-ventricular orifice* is seen, guarded by two membranous curtains (*bicuspid*, or *mitral valve*) of unequal size, that next the aortic orifice being the larger. From the edges of these tendinous chords (*chordæ tendineæ*) pass to the muscular pillars (*columnæ papillares*) upon the walls of the ventricle.

f. Note the size of the left ventricular cavity and the great thickness of its walls.

128. Make an opening in the left auricular wall opposite the point of entrance of the pulmonary vein, and note—

a. The thinness of its walls when compared with the ventricle.

b. The peculiar arrangement of its muscles (*musculi pectinati*).

c The *pulmonary veins* entering it on its inner side.

d. The musculo-membranous valve, stretching from one wall of the auricle to the other.

129. Trace the *pulmonary arteries* until they are found to enter the right ventricle by a common trunk. Pass the scissors through this into the right ventricle and slit it open along the ventral side of the heart a little to the right of the incision made in opening the left ventricle, and note

    a. That the *pulmonary orifice* is guarded in the same manner as the aortic, by *semilunar valves*, which, however, are more delicate though similar in structure and arrangement. Behind them are also found pouches (*sinuses of Valsalva*).

    b. The right *auriculo-ventricular orifice* is guarded by a large fleshy curtain on the side farthest from the pulmonary opening, and on the side next it by a small musculo-membranous one. They arise from the dorsal wall of the ventricle, and their edges meeting below have a common attachment to the ventral wall. Note that the walls of the right ventricle are much thinner than those of the left, and not marked by *columnæ papillares*.

    c. *Open the right auricle* and search carefully for the open mouths of the veins of the heart which empty into it.

    d. Note the entrance of the *superior* and *posterior cavæ*; also, on its inner side a valve-like arrangement of the muscles, similar to that seen on the left side.

130. Dissection of the Brain, Spinal Cord, and Cranial Nerves.—Cut away with a sharp scalpel the vault of the cranium from the foramen magnum to the anterior nares, and observe that the brain, the upper surface of which is now exposed, fills the skull cavity almost completely. After removing the pia mater, note from before, backward, the following:

THE BRAIN.   151

a. The *olfactory lobes* (*rhinencephalon*), two slender grayish prolongations from the hemispheres, which, lying at first side by side, pass forward toward the nares, just above and to the outside of the septum of the orbits.

b. The *cerebral hemispheres* (*prosencephalon*), smooth and joined in front in the median line, but separated behind by a deep cleft.

c. Carefully separate the cerebral hemispheres posteriorly, and lying between them, and in front of the cerebellum, note the small red *pineal gland*, or *epiphysis cerebri*, which rests on the roof of the third ventricle.

d. The *cerebellum*, longer in its antero-posterior diameter than in its transverse, convex from side to side, and from before backward, deeply convoluted transversely, and having on either side a process (*flocculus*) lodged in a cavity of the skull, which is arched over by the anterior vertical semi-circular canal.

e. Outside the cerebellum the *optic lobes* (*corpora bigemina*), two round white bodies which seem to be pushed aside by the increased development of the former. They are lodged in large cavities in the side of the cranial box, and are separated from the hemispheres above by a sort of *tentorium*.

f. Behind the cerebellum (which must be raised to see it) the *medulla oblongata*, having on its dorsal side the—

g. *Fourth ventricle;* the floor of which is distinguished by its gray color. Behind, it is bounded by two rounded white eminences (*restiform bodies*), which are the continuations of the *posterior columns* of the cord, and pass in front into the cerebellum, of which they constitute the *posterior peduncles*.

 h. Note the commencement of the canal (*iter a tertio ad quartum ventriculum*, or *aqueduct of Sylvius*) which connects the fourth with the third ventricle. This may be seen by gently parting the brain substance in the middle line at the anterior portion of the fourth ventricle, and turning the cerebellum forward.

 i. Also the ventricle in the substance of the cerebellum which communicates with the fourth ventricle.

 j. The rounded mass of gray matter which forms part of the roof of the fourth ventricle.

 k. Observe that the brain is placed at a right angle to the spinal cord, which is sharply bent where it joins the medulla.

131. Raising the brain a little and pressing it gently to the right, note—

 a. The large thick *optic nerve* entering the orbit.

 b. Appearing upon the upper surface of the left crus of the brain behind the optic lobe, and passing forward under and a little to the outer side of the optic nerve to enter the orbit, the *fourth nerve* (*trochlear*). Trace its origin till it is found to arise from the roof of the fourth ventricle (*valve of Vieussens*) in the median line.

 c. Raising the brain still more, the *third nerve* (*motor oculi*) will be seen arising from the inner and under surface of the crus and almost immediately passing out of the skull into the orbit beneath the optic.

d. Internal to the third pair of nerves note the *infundibulum*, a prolongation of gray matter from the floor of the third ventricle.

e. Attached to the ventral end of the infundibulum the *pituitary body*, or *hypophysis cerebri*, which is lodged in the *sella turcica*, and will probably be torn away when the brain is removed from the skull.

f. Arising from the side of the medulla the *fifth nerve*, having on it the *Gasserian ganglion*, lodged in a hollow of the bone (22, *e*).

g. Behind the fifth the *seventh nerve* leaves the side of the medulla and enters a foramen seen in the lateral wall of the skull (22, *d*).

h. Posterior and a little dorsal to the seventh may be seen the *eighth*, which, arising from the medulla, enters the auditory capsule (22, *d*). Its origin may readily be traced to the floor of the fourth ventricle.

i. Arising from the under surface of the anterior portion of the medulla near the median line the *sixth nerve* (*abducens*), which immediately enters the floor of the skull (22, *f*).

j. Behind the eighth nerve the *ninth* (*glossopharyngeal*), *tenth* (*pneumogastric*), and *eleventh* (*spinal accessory*) nerves arise from the medulla and pass out of the skull through a foramen situated beneath the foramen for the eighth nerve (22, *c*).

k. Arising from the under surface of the posterior part of the medulla the *twelfth* nerve

7*

(*hypoglossal*), which leaves the skull by a foramen in the exoccipital bone (22, *b*).

l. Between the skull and atlas is seen passing out the *sub-occipital* or *first cervical* nerve, which is very large.

132. Cutting the nerves arising from the brain and removing the latter from before backward, note on its base—

a. In front the *olfactory lobes* (*rhinencephalon*).

b. The *optic chiasma*, which, if great care is not taken to cut very close to the anterior wall of the skull, will be damaged, as the optic nerves pass out of the brain-case into the orbit as soon as they divide.

c. Behind the chiasma the *infundibulum*, a gray projection which forms part of the floor of the third ventricle (133, *d*).

d. On either side of the infundibulum the *crura of the cerebrum*, which are continuous with the anterior columns of the cord and ventral portion of the medulla.

e. To the outer side of these the *optic*, or *mesencephalic lobes*.

f. Arising from the inner side of the crura the *third pair* of nerves (141, 143, 144).

g. Behind the crura and continuous with them the *medulla oblongata*, which is itself continuous with the cord.

h. Observe the origin of the *sixth pair* from the under surface of the medulla near its anterior part (142).

i. The origin of the *twelfth pair* (148) will now

THE BRAIN. 155

be more distinctly seen at the posterior portion of the medulla near the median ventral line.

j. Note the absence of a *pons Varolii*; this is correlated with the absence of the lateral lobes of the cerebellum.

133. Cut through the spinal cord at the junction of the skull and atlas, and remove the brain. Place it in alcohol to harden, and then make a median anteroposterior vertical section of it with a sharp razor, first parting the hemispheres, which behind are only united by connective tissue, and note—

a. The continuity of the fourth ventricle with the central canal of the cord.

b. The cut surface of the cerebellum showing the ventricle which extends into its substance and communicates with the fourth ventricle; and the arrangement of its convolutions.

c. The *aqueduct of Sylvius*, which places the fourth ventricle in communication with the third.

d. The *third ventricle*, bounded laterally by the *optic thalami* and communicating in front on either side through an opening, the *foramen of Munroe*, with the *lateral* (*first* and *second*) *ventricles*.

e. Bounding the third ventricle anteriorly is the *lamina terminalis*, a thin sheet of nervous matter which connects the cerebral hemispheres. Below, the third ventricle is continued downward into the infundibulum.

f. By drawing the hemisphere away from the rest of the brain an excellent view may be had of the fibres of the crus of the same side (134, *d*), becoming continuous with the substance of the cerebral hemisphere.

g. Note that the optic lobes have ventricles and that there is no *corpus callosum*. Make sections of the brain substance in different planes parallel to the base, beginning on its dorsal surface, and note the distribution of white and gray matter.

134. With a pair of curved scissors cut away the neural arches of all the vertebræ, taking care not to lacerate the spinal cord, and note—

a. That in the cervical region the cord is of almost uniform size until near the base of the neck, where the *cervical enlargement* may be seen, from which the *brachial plexuses* arise (69).

b. That there is a second enlargement, opposite the sacrum, from which the sacral and lumbar plexuses arise (153, 154). The posterior columns are here separated, leaving a sort of ventricle between them (*sinus rhomboidalis*), which communicates in front and behind with the central canal of the cord.

c. Behind the sinus the cord becomes very small until it is scarcely visible in the posterior caudal region. There is no *cauda equina*, the cord itself being continued to the extreme end of the spinal column, even into the *pygostyle*.

135. Examine one of the spinal nerves, and observe that it is formed by the union of an anterior and posterior root, each of which arises from the spinal cord. The union of the roots takes place almost as soon as they leave the cord, and the common trunk thus formed has on it a ganglion at the point at which it leaves the spinal canal. If one of the nerves of the cervical plexus is examined, the ganglion will be found upon the posterior root.

136. Remove the spinal cord, and note on its ventral surface the anterior median fissure. Then place it in alcohol to harden, and make sections of it at various levels, and note—

a. The central mass of gray matter which, faintly seen in some parts, becomes quite distinct in the regions of the cervical and lumbar enlargements. In the dorsal region it is also quite distinct.

b. The distinctly marked *anterior fissure*.

c. Dorsal to the anterior fissure the *canalis centralis*, a minute opening which may be seen on examining the cut surface of the cord with a hand lens.

137. Distribution of the Cranial Nerves.—The *olfactory nerve* overlies the roof of the nostril, and is distributed to its mucous membrane (130, *a*).

138. Chip away the inner wall of the orbit down to the point at which it is pierced by the *fourth nerve* (131, *b*), which passing over the superior rectus muscle (155, *b*) near its origin, runs forward along the inner wall of the orbit to be distributed to the superior oblique muscle (155, *a*).

139. The *optic* may be seen to enter the inner side of the eyeball at its posterior internal part, by drawing the latter outward.

140. Remove the bone so as to expose the *Gasserian ganglion* of the fifth, a small reddish body which lies in a cavity in front of the ridge which bounds anteriorly the fossa for the medulla oblongata; it will be found to give off three branches, viz.:

    a. The **Ophthalmic Nerve**, which runs forward through the floor of the skull, passes beneath the origin of the superior rectus (155, *b*) and superior oblique (155, *a*) muscles, and enters the nasal cavity, along the inner side of which it may be traced; and if great care be taken the branch can be followed to the tip of the upper mandible, where it is distributed beneath its horny covering.

    b. The Superior Maxillary **Nerve** passes outward, and, after giving branches to the skin of the eyelids and mucous membrane of the eye (*conjunctiva*), runs directly beneath the globe of the eye, and can be traced to the outer side of the lachrymal bone, where it breaks up and is distributed to the skin, etc., in this region.

    c. The **Inferior** Maxillary **Nerve**, the largest of the division of the fifth, passing downward and then forward beneath the eyeball, divides into two branches, one of which supplies the muscles of mastication, to which it is distributed; the other runs forward and enters the inferior maxillary bone by a foramen on its inner side, and then again emerges on its outside beneath the horny covering, where it is distributed.

141. Chip away a little more bone so as to reach the exit of the *third nerve* (131, *c*). Cut through the attachment of the superior rectus muscle (155, *b*), and turn it inward so as to expose its under surface; it will be seen to be supplied by branches from the third nerve which enter it beneath. The nerve then passes beneath the optic nerve.

142. To the outer side of the main trunk of the third nerve may be seen the *sixth nerve* (131, *i*), which is almost immediately distributed to the internal rectus muscle (155, *f*). Trace it back through the floor of the skull to the point where it was seen to enter the latter after leaving the brain.

143. Now return to the third nerve, and follow it as it crosses below the origin of the inferior rectus muscle, which it supplies (155, *d*). It then sends a large branch to the inferior oblique (155, *e*), and a smaller one to the internal rectus (155, *f*), and terminates in the *Harderian gland* (155, *g*).

144. As it passes behind the optic nerve it is in close relation to a small ganglion (*ophthalmic ganglion*), which in turn sends a pair of branches forward which pierce the *sclerotic* on opposite sides of the optic nerve near its entrance (*ciliary nerves*).

145. Stuff the mouth and upper portion of the œsophagus with cotton. Divide the œsophagus and trachea about an inch below the head, and turn them forward after carefully dissecting them from the parts above them; then divide the jugular veins and carotid arteries, and turn them forward with their branches, and note—.

    a. The common trunk of the *pneumogastric* and *spinal accessory* (*tenth* and *eleventh cranial nerves*), which should be traced to the base of the skull, which it will be found to leave by a foramen common to it and the glosso-pharyngeal.

    b. The pneumogastric just behind the articulation of the lower jaw with the skull gives off the superior laryngeal, which runs forward along the inner side of the horn of the hyoid to reach the trachea, upon which it divides into an anterior and a posterior branch; the former is distributed to the larynx, the latter supplies the trachea.

146. Directly in front of the point of exit of the pneumogastric is a *large ganglion*, situated on the glosso-pharyngeal nerve just where it emerges from the skull. The glosso-pharyngeal or ninth cranial nerve, after leaving this ganglion, which also sends off the sympathetic trunk, passes downward and on the inner side of the posterior portion of the horn of the hyoid, divides into two branches, one of which, the *descendens noni* (73), runs down the neck; the other, accompanied by a branch of the carotid artery, runs along the inner side of the horn of the hyoid, and supplies the pharynx and tongue.

147. Immediately outside of the glosso-pharyngeal is the *seventh cranial nerve*, which may be seen, by pushing its ganglion toward the median line, to pass out from the skull by a foramen, behind that for the carotid artery (19, *l*). It then turns outward behind the articulation of the lower mandible, and may be followed as it passes over the muscles of mastication to the side of the skull.

148. Passing out of the skull by the condyloid foramen (19, *m*) is the *hypoglossal* (*twelfth cranial*) nerve (132, *i*), which, lying on the ventral surface of the origin of the neck muscles, passes outward above the pneumogastric trunk, and then turns forward and runs along the outer side of the horn of the hyoid until it gets opposite the larynx, where it divides into two branches, one of which runs down the trachea ; the other passes forward to the tongue.

149. The Sympathetic Nerve.—Remove one lung, find the sympathetic nerve in the dorsal region, and from this point work forward until the base of the skull is reached, and backward until the nerve is lost upon the bodies of the movable caudal vertebræ. The nerve will be found to start from a ganglion (146) at the base of the skull, and to run backward along the sides of the centra of the cervical vertebra, passing through the foramina in their transverse processes, accompanied by the vertebral artery and vein.

150. After passing ventral to the brachial plexus, it runs ventral to all the ribs, its ganglia being also connected in the dorsal region by commissures passing between the necks of the ribs and their tubercular transverse processes. In the lumbar and sacral regions the cord lies ventral to the sacral and lumbar nerves, and, after passing dorsal to the large transverse process of the first caudal vertebra, continues backward ventral to the caudal spinal nerves until it meets the sympathetic chain of the opposite side, beneath the centrum of the second movable caudal vertebra.

151. In the cervical region the sympathetic is intimately connected with all the spinal ganglia except that of the suboccipital nerve. In the dorsal region it has a ganglion partly fused with that of each dorsal spinal nerve. In the lumbar and sacral regions the ganglia are very indistinct, and absent in the caudal region. Branches are given off from the ganglia between the first and second, second and third, third and fourth, fourth and fifth dorsal vertebræ, which unite to form the splanchnic nerve, and from which branches run to a plexus on either side of the dorsal aorta (*solar plexus*) (100), at the point where the cœliac axis and superior mesenteric arteries are given off, both of which are accompanied by trunks from this source.

152. Spinal Nerves.—The *cervical nerves* and *brachial plexus* have been already examined. There are five dorsal nerves. The last cervical and four of the dorsal supply the thoracic walls, the last dorsal being distributed to the abdominal wall.

153. The *lumbar* nerves, three in number, unite to form a *plexus* from which trunks are given off to the inner and anterior aspect of the thigh (*crural* and *obturator nerves*).

154. The next four nerves are the sacral, and unite

THE MUSCLES OF THE EYE.

to form the *sciatic nerve*, which, passing backward and outward, issues from the pelvis by the ilio-sciatic foramen (48), and runs down the inner posterior aspect of the thigh to supply the hind limb (79). Behind, the caudal vertebræ nerves pass out to supply the various structures of the pelvis and uropygium.

155. The Muscles of the Eye.—Having cut away the inner wall of the orbit as in the dissection of the cranial nerves (137), note—

- *a.* The *superior oblique* arising from the anterior inferior portion of the inner wall of the orbit, and running backward and outward to be attached to the upper and outer portion of the eyeball.
- *b.* The *superior rectus* muscle arises from the orbital wall behind the optic nerve and runs outward and a little forward to be attached to the upper and outer portion of the globe.
- *c.* The *external rectus* arises just behind the last muscle and is attached to the globe posteriorly near its outer edge.
- *d.* The *inferior rectus* arises from the inner wall of the orbit just below the entrance of the optic nerve, and running outward and a little forward is attached to the outer portion of the under surface of the eyeball.
- *e.* The *inferior oblique* muscle runs from the anterior inferior part of the wall of the orbit, and passing outward and backward is attached to the eyeball external to the outer extremity of the inferior rectus.
- *f.* The *internal rectus* arises in front of the optic

nerve and runs forward to be attached to the eyeball at its anterior margin.

g. Just beneath the inferior oblique lies the **Harderian gland**, a flattened, bean-shaped body, paler in color than the muscles.

h. Arising from the sclerotic in front of the optic nerve entrance a small muscle (*pyramidalis*), running backward and upward and terminating in a tendon which turns around the optic nerve and runs downward and forward and then upward to be attached to the free border of the nictitating membrane which is drawn forward when the muscle contracts.

i. A broad *bursalis* muscle, which, arising from the posterior superior portion of the sclerotic, runs forward and downward, and terminates in a fibrous sheath enclosing the tendon of the *pyramidalis*. The *bursalis* contracting at the same time as the *pyramidalis*, prevents the tendon of the latter from exercising injurious pressure upon the optic nerve.

156. The Eyeball.—Remove the eyeball, leaving attached to it as much of the optic nerve as possible. Clear away the pyramidalis muscle, and note the peculiar *turnip-shape* of the eyeball; the *optic nerve* entering the eye on its inner side, below its axis; the opaque *sclerotic;* the *cornea*, transparent and more sharply curved than the sclerotic, beyond which it projects considerably; the *conjunctiva*, lining the inner surfaces of the eyelids and reflected over the front of the eyeball; the *iris*, a gold-colored ring, having a narrow border internally of black surrounding the circular pupil.

157. Holding the eyeball between the forefinger and thumb, with a pair of small scissors cut through the sclerotic along the most prominent part of the ridge, which marks the union of its flatter and more rounded parts, and note—

- a. The *two ciliary nerves* running over the outer surface of the choroid close together from the optic nerve region to the iris.
- b. The *choroid*, dark and deeply pigmented, lining the sclerotic from the optic nerve entrance to the outer border of the iris.
- c. Divide the choroid along the same line as the sclerotic was divided, and note that the cavity of the globe is filled by the transparent gelatinous *vitreous humor*, which should be removed.
- d. The choroid will now be seen to be lined by a delicate pearly membrane, the *retina*, which is thickest at the inner portion of the eye, thinning away as it approaches the iris.
- e. The *pecten*, a brown vascular fringe, projects from the choroid, near the optic nerve entrance, into the vitreous.
- f. The *lens* closes the opening of the pupil behind and is held in place by its suspensory ligament, which is attached about the outer margin of the iris; remove it and observe that its outer surface is less convex than its deeper.
- g. The *iris* may now be seen from behind, and outside of it a great number of fine plaits radiating from its margin; these are the *ciliary processes*. Tear away the iris and it will be seen to separate readily from the choroid

about opposite the junction of the cornea and sclerotic.

h. Scrape away the ciliary processes and a *bony ring* will be exposed surrounding the cornea and forming the boundary of the sclerotic.

158. The Ear.—Carefully cut away the postero-lateral portion of the outer table of the skull and the cancellated bone tissue beneath it (*diploë*), and note—

a. The *anterior vertical semi-circular canal* (the longest of the three bony canals forming part of the *labyrinth*), which lies higher on the side of the cranium and reaches farther back than the others, and lies in a vertical plane which is directed backward and inward. Its anterior end is dilated, forming the *ampulla*, which communicates with the anterior superior portion of the vestibule, and its posterior extremity unites with that of the *posterior vertical semi-circular canal*.

b. The *posterior vertical semi-circular canal* lies in a nearly vertical plane directed from without inward and forward. Its superior end enters the vestibule above, and its lower extremity is marked by an *ampulla* which communicates with the vestibule below.

c. The *external horizontal canal* lies in a horizontal plane, and communicates in front with the vestibule through the medium of its ampullary extremity, and behind by its non-ampullary end.

d. The *vestibule* is a small chamber which communicates, as just seen, with the five open-

THE EAR.

ings of the semi-circular canals, and anteriorly communicates with the *cochlea*, which presents *no trace* of spiral structure.

e. Note the *fenestra ovalis*, which is filled by the expanded end of the *columella auris*.

f. The *columella* is a very short minute rod of bone attached to the inner side of the membrana tympani by its outer end, and having its inner end expanded and fixed as just seen in the fenestra ovalis of the vestibule (19, *g*).

159. Cut away the walls of the external auditory meatus in a fresh specimen (noticing that it is lined by a reflection of the skin) and expose the drum membrane, through which may be seen the anterior extremity of the columella, which is attached to it.

160. Remove the membrana tympani, which divides the external auditory meatus from the tympanum, taking care not to displace the columella, which will now be seen to pass backward through a canal at the posterior part of the floor of the middle ear. In front of the canal for the columella is a large cavity communicating by an opening in the outer part of its floor with the Eustachian canal (19, *i*), and internally with the air spaces in the cancellated structure of the basi-sphenoid. Pass a bristle to prove this. Behind the canal for the columella is a second opening communicating with the diploë of the sides and roof of the cranium. The tympanum, the canal for the columella, and the Eustachian canal are lined by mucous membrane.

INDEX.

ABDUCENS (sixth cranial) nerve, 153, 154, 159
Acetabulum, 122, 123
Aditus laryngis, 102, 142
Adrenal bodies, 147
Air sacs, 126, 127, 133, 146; abdominal, 133, 143; thoracic, 133
Alisphenoids, 110
Alula, 99
Ampulla, 166
Angulare, 113
Antebrachium, 94, 99, 101
Anterior abdominal vein, 137; nares, 105; tibial arteries, veins, and nerves, 132; vertical semicircular canal, 166
Antitrochanter, 123
Aorta, 135, 148; dorsal, 135, 137, 139, 140
Apteria, 98
Aqueduct of Sylvius, 152, 155
Arm, 94
Artery, anterior tibial, 132; ascending cervical, 135; brachial, 129, 136; bronchial, 139; common carotid, 114, 128, 130, 135; coronary of heart, 149; external carotid, 131; femoral, 132; iliac, 140; inferior mesenteric, 140; innominate, 128, 135; intercostal, 139; internal carotid, 131; pectoral, 129, 136; posterior tibial, 132; pulmonary, 138; sciatic, 140; subclavian, 128, 136; superior mesenteric, 140; vertebral, 114, 135
Arytenoid cartilages, 142
Atlas, 115
Auditory meatus, 94, 103, 167; eighth cranial nerve, 153

Auricles of the heart, 134, 136, 148, 149, 150
Auriculo-ventricular orifice, 149, 150
Axis, 115

BARBICELS, 97
Barbs, 96
Barbules, 96
Basi-occipital bone, 110, 111
Basi-sphenoid bone, 103, 109
Beak, 105
Bicuspid valve, 149
Bill, 93, 105
Bone, alisphenoid, 110; angulare, 116; articulare, 113; dentary, 113; ethmoid, 107; exoccipitals, 110, 111; frontal, 103; hyoid, 113; ilium, 122; ischium, 123; jugal, 106; lachrymal, 105; maxillary, 106; nasal, 107; palatine, 105; parietal, 109; premaxillary, 107; presphenoid, 108; pterygoid, 106; quadrate, 106; quadrato-jugal, 107; splenial, 113; supra-occipital, 110; surangulare, 112
Bones, carpal, 121; metacarpal, 122; of fore-limb, 120; of hind-limb, 123; of mandible, 112; of pectoral arch, 119; of pelvic arch, 122; of skull, 105; of spinal column, 113; tarso-metatarsus, 124; tibio-tarsus, 124
Bony and cartilaginous skeleton, 102; ring of sclerotic, 165.
Brachial arteries, 129; nerve, 129; plexus, 127, 129, 156
Brachium, 101
Brain, 150

Bronchi, 141
Bursa Fabricii, 148
Bursalis muscle, 164

CALAMUS, 95

Canalis centralis, 157
Canals, carotid, 104; Eustachian, 103; semicircular of ear, 111; vertebral, 135
Capitulum, 118
Cardiac opening of stomach, 145
Carotid arteries, 114, 128, 130, 135
Carpal bones, 121
Cartilage, arytenoid, 142; Meckel's, 112; naso-ethmoidal, 108; thyroid, 143
Cauda equina, 156
Caudal spinal nerves, 163; vertebræ, 117
Cava, inferior, 136, 137, 150; posterior, 137; superior, 137
Centra of cervical vertebræ, 113; of dorsal vertebræ, 116; of lumbar vertebræ, 116
Cerebellum, 151, 152, 155
Cerebral hemisphere (prosencephalon), 111, 151
Cervical enlargement, 156; spinal nerves, 130; vertebræ, 113
Characters, zoölogical, of Aves, as distinguished from Reptilia, 90; of Sauropsida, 89; of Schizognathæ, 92
Chordæ tendineæ, 149
Choroid, 165
Ciliary process, 165
Circumflex nerve, 129
Clavicle, 120
Claws, 95
Cloaca, 148; opening of, 101
Cnemial process, 124
Cochlea, 167
Cœco-colic diverticula, 144
Cœliac axis, 139
Colon, 144
Columella auris, 104, 167
Columnæ papillares, 149, 150
Condyles of femur, 124; of humerus, 121
Condyloid foramen of skull, 104, 111, 161
Conjunctiva, 164
Contour feathers, 97
Coracoid, 119

Cornea, 164
Coronary arteries, 149
Corpora bigemina, 151
Corpus Arantii, 149; callosum, 156
Coverts, tail, 100; wing, 99
Cranial nerves, 150; distribution of, 157
Crop, 100, 126
Crura of cerebrum, 154, 155
Crural nerves, 162
Crus, 94, 101

DENTARY, 113

Descendens noni nerve, 130, 160
Diaphragm, 146
Digits of manus, 101, 122; of pes, 125
Diploë, 166
Dissection of brain, spinal cord, and cranial nerves, 150; of heart, 148; of pleuro-peritoneal cavity, 132
Distribution of cranial nerves, 157
Dorsal aorta, 135, 137, 139; spinal nerves, 162; vertebræ, 116
Down feathers, 97
Drum membrane, 167
Duct of pancreas, 145
Duodenum, 144

EAR, 166

Eighth cranial (auditory nerve), 153
Eleventh cranial (spinal accessory) nerve, 153, 160
Epidermic exoskeleton, 95
Epignathous, 94
Epiphysis cerebri, 151
Ethmoid, 107
Eustachian canal, 103, 104, 109, 167
Exoccipital bones, 110, 111
External appearance, 93; auditory meatus, 94, 103, 167; external carotid arteries, 131; rectus muscle, 163; syringeal muscles, 126, 129, 141
Eye, 94
Eyeball, 164
Eyelids, 94

FACIAL (seventh cranial) nerve, 153, 160

Fallopian tubes, 148

INDEX. 171

Feathers, 93, 95; around external auditory meatus, 94; arrangement of, 98; contour, 97; down, 97; primaries, 98; secondaries, 98; structure of, 95; tail, 99; wing, 98
Femoral artery, 132; vein, 132, 137
Femur, 101, 123; condyles, 124
Fenestra ovalis, 167
Fibula, 124
Fifth cranial (trigeminal) nerve, 104, 153
Filo-plumæ, 97
First cranial nerve (olfactory), 157
Fissure of spinal cord, 157
Flocculus, 151
Foot, 94
Foramen, condyloid, 161; ilio-sciatic, 123, 163; magnum, 104; obturator, 123; of Monro, 155; ovale, 104
Foramina of exit of cranial nerves, 104, 111
Forearm, 94
Fore-limb, 101
Fossa for cerebellum, 111; for Gasserian ganglion, 111; glenoid, 120
Fourth cranial (trochlear) nerve, 152, 157; ventricle, 151, 153, 155, 157
Frontal bones, 108
Furculum, 120

Gall-bladder and ducts, 144
Ganglion of glosso-pharyngeal nerve, 160; ophthalmic, 159
Ganglion, spinal, 157; sympathetic, 162
Gasserian ganglion, 111, 153, 158
Gizzard (gigerium), 135, 139, 140, 143, 145
Gland, Harderian, 164, 159; pineal, 151; thymus, 136
Glenoid fossa, 120
Glosso-pharyngeal (ninth cranial) nerve, 153, 160; ganglion of, 160
Greater tuberosity of humerus, 121
Great omentum, 144
Great trochanter of femur, 123

Hallux, 125
Harderian gland, 159, 164
Head, 100

Heart, 134, 148
Hemisphere, cerebral, 111
Hind-limbs, 94
Humerus, 120; condyles of, 121
Hyoid bone, 113
Hypocleidium, 120
Hypoglossal (twelfth cranial) nerve, 153, 154, 161
Hypophysis cerebri, 153

Iliac artery, 140
Ilio-sciatic foramen, 123, 163
Ilium, 122
Inferior cava, 136, 137, 150; maxillary nerve, 158; mesenteric artery, 140; oblique muscle, 159, 163; rectus muscle, 159, 163
Infundibulum, 153, 154
Innominate artery, 128, 135; vein, 129
Intercostal arteries, 139
Interior of mouth, 101
Internal carotid arteries, 131; rectus muscle, 159, 163
Interorbital septum, 105
Intervertebral disks, 131
Intestine, small, 135, 144
Iris, 94, 164, 165
Ischium, 123

Jugal bone, 104, 106
Jugular vein, 129, 130

Keel of sternum, 119
Kidneys, 147

Lachrymal bone, 105
Lachrymo-nasal opening, 105
Lamina terminalis, 155
Larynx, 142
Lateral (first and second) ventricles, 155
Left pneumogastric nerve, 138; superior cava, 136
Lens of the eye, 165
Ligament, transverse, 115
Ligamentum nuchæ, 131

Liver, 134, 140, 144
Lobes, olfactory, 151, 154; optic, 151, 154
Lophosteon, 119
Lumbar plexus, 162; spinal nerves, 162; vertebræ, 116
Lungs, 133, 146

M ANDIBLE, 93, 112

Manus, 94, 101
Maxilla, 103, 106
Meckel's cartilage, 112, 113
Medulla oblongata, 151, 154
Membrana, tympani, 141, 167; tympaniformis interna, 141; nictitans, 94; semilunaris, 141
Mesentery, 144
Metacarpal bones, 122
Metatarsal bones, 125
Metostea, 119
Mitral valve, 149
Mobility of upper mandible, 94
Motor oculi (third cranial) nerve, 152, 154
Muscle, bursalis, 164; external rectus, 163; external syringeal, 126, 129, 141; inferior oblique, 159, 163; inferior rectus, 159, 163; internal rectus, 159, 163; laryngeal, 142; pectoral, 127, 128; perching, 132; pyramidalis, 164; superior oblique, 163; superior rectus, 159, 162
Musculi pectinati, 149

N ASAL bones, 107; cavity, 101, 112

Naso-ethmoidal cartilage, 108
Neck, 100
Nerves, anterior tibial, 132; brachial, 129; caudal spinal, 163; cervical spinal, 130; ciliary, 159, 165; circumflex, 129; crural, 162; descendens noni, 130, 160; dorsal spinal, 162; eighth cranial (auditory), 153; eleventh cranial (spinal accessory), 153, 160; exit of cranial, 104, 111; fifth cranial (trigeminal), 104, 153; first cranial (olfactory), 157; fourth cranial (trochlear), 152, 157; inferior maxillary, 158; lumbar spinal, 162; ninth cranial (glossopharyngeal), 153, 160; obturator, 162; of brachial plexus, 127, 129; ophthalmic, 158; posterior tibial, 132; recurrent laryngeal, 138; sacral, 162; sciatic, 132, 163; second cranial (optic), 152, 164; seventh cranial (facial), 153, 154, 160; sixth cranial (abducens), 153, 154, 159; splanchnic, 162; superior laryngeal, 160; superior maxillary, 158; sympathetic, 160, 161; tenth cranial (pneumogastric), 129, 138, 139, 153, 160; third cranial (motor oculi), 152, 154, 159; twelfth cranial (hypoglossal), 153, 154, 161
Neural arch of cervical vetebræ, 113, 114
Nictitating membrane, 94, 164

O BTURATOR foramen, 123; nerves, 162

Occipital condyle, 104
Odontoid process, 115
Œsophagus, 126, 143, 145
Olfactory lobes (rhinencephalon), 151, 154; nerves (first cranial), 157
Ophthalamic ganglion, 159; nerve, 112, 158
Optic chiasma, 154; lobes, 151, 154, 156; nerve, 152, 154, 158, 164; thalami, 155
Orbito-temporal fossa, 104
Os angulare, 113; articulare, 113
Ovary, 147
Oviduct, 147

P ANCREAS, 135, 140, 144; duct of, 145

Palatine bones, 105; processes of maxilla, 103
Parietal bones, 109
Patella, 124
Pecten, 165
Pectoral artery, 129, 136
Pectoralis major muscle, 127, 128; minor muscle, 127, 128
Peculiar cervical vertebræ, 115
Pelvic girdle, 122
Penna, 97
Perching muscle, 132
Pericardium, 134
Periotic capsule, 109

INDEX.

Pes, 95, 101
Pineal gland, 151
Pituitary body, 153
Pleurostea, 119
Plumulæ, 97
Pneumatic openings in bones, 121, 133
Pneumogastric (tenth cranial nerve), 129, 138, 139, 153, 160
Pollex, 101, 122
Pons Varolii, 155
Portal vein, 137
Posterior cava, 136, 137, 150; clinoid processes, 109; columns of cord, 152; nostril, 101, 103; peduncles, 152; tibial artery, 132; tibial nerve, 132; tibial vein, 132
Postzygapophyses, 114
Premaxilla, 107
Presphenoid, 108
Prezygapophyses, 114
Process, cnemial, 124; odontoid, 115; spinous, 116, 117; transverse, 116, 117; tubercular of vertebræ, 114, 116; uncinate, 118; xiphoid, 118
Proscencephalon (cerebral hemisphere), 151
Proventriculus, 143, 145
Pterygoids, 103, 106
Pterylæ, 98
Ptilosis, 98
Pubis, 123
Pulmonary artery, 138, 149; vein, 139, 149
Pygostyle, 117, 156
Pylorus, 145
Pyramidalis muscle, 164

QUADRATE bones, 103, 106
Quadrato-jugal, 104, 107

RACHIS, 95
Radius, 121
Rectrices, 99
Rectum, 148
Recurrent laryngeal nerves, 138
Remiges, 100
Reproductive apparatus in female, 147; in male, 146
Restiform bodies, 152
Retina, 165

Rhinencephalon (olfactory lobes), 151
Ribs, 117; rudimentary, 114; vertebral, 117; sternal, 118
Right pneumogastric nerve, 138, 139; superior cava, 137
Roots of spinal nerves, 157
Rostrum of sphenoid, 101, 103; of sternum, 118
Rudimentary ribs, 114

SACRAL enlargement, 156; nerves, 162; vertebræ, 116
Scapula, 120
Scapus, 95
Sciatic artery, 140; nerve, 132, 163
Sclerotic, 159, 164
Scutella, 95
Second cranial (optic) nerve, 152, 164
Sella turcica, 109, 153
Semicircular canals, 111, 160
Semilunar valves, 148, 150
Semiplumæ, 97
Seventh cranial (facial) nerve, 153, 160
Shoulder girdle, 119
Sinuses of Valsalva, 148, 150
Sinus rhomboidalis, 156
Sixth cranial (abducens) nerve, 153, 154, 159
Skull, 102
Small intestine, 135, 144
Solar plexus, 140, 162
Spermatozoa, 147
Spinal accessory (eleventh cranial) nerve, 153, 160; cord, 150, 156, 157; ganglia, 157; nerves, 130, 162
Splanchnic nerve, 162
Spleen, 139, 144
Splenial, 113
Squamosal bone, 108
Sternum, 118
Stomach, 135, 143, 145
Subclavian artery, 128, 136
Suboccipital (first cervical) nerve, 154
Suborbital bony bar, 104, 107
Superior cava, 137, 150; laryngeal nerve, 160; maxillary nerve, 158; mesenteric artery, 140; oblique muscle, 163; rectus muscle, 159, 163
Supra-occipital bone, 110

Surangulare, 112
Sympathetic nerve, 114, 160, 161
Syringeal muscles, 126, 129, 141
Syrinx, 141

TARSO-METATARSUS, 94, 95, 101, 124
Tenth cranial (pneumogastric) nerve, 129, 153, 160
Tentorium, 151
Testes, 146
Thigh, 94
Third cranial (motor oculi) nerve, 152, 154, 159; ventricle, 155
Thoracic air-sacs, 133
Thymus gland, 136
Thyroid cartilage, 143; gland, 129, 143
Tibio-tarsus, 124
Tongue, 102
Trachea, 126, 142
Transverse ligament, 115; processes of vertebræ, 116, 117
Trigeminal (fifth cranial) nerve, 104, 153
Trochlear nerve, 152
Trunk, 100
Tuberculum, 118
Tuberosity, greater, of humerus, 121
Twelfth cranial (hypoglossal) nerve, 153, 154, 161
Tympanum of ear, 167; of syrinx, 141

UMBILICUS of feather, inferior, 95; superior, 96
Ulna, 121
Ureter, 147

Uropygial gland, 100
Uropygium, 100, 117

VALVE, bicuspid, 149; musculo-membranous of auricles, 149, 150; of right auriculo-ventricular orifice, 150; of Vieussens, 152; semilunar, 148-150
Vanes, 96
Vas deferens, 146, 148
Vein, anterior abdominal, 137; anterior tibial, 132; femoral, 132, 137; jugular, 129, 130; left innominate, 129; left superior cava, 136; portal, 137; posterior cava, 136, 137, 149; posterior tibial, 132; right superior cava, 137
Ventricle, first and second, 155; fourth, 151, 153, 155, 157; in substance of cerebellum, 152; third, 155
Ventricles of heart, 134, 148, 149
Vertebræ, caudal, 117; cervical, 113; dorsal, 116; lumbar, 116; sacral, 116
Vertebral artery, 114, 135; canal, 114, 135; vein, 114
Vestibule of ear, 166
Vitreous humor, 165
Vocal organs, 141
Vomer, 109

WINGS, 93, 94

Wing coverts, 99

XIPHOID processes, 118

www.ingramcontent.com/pod-product-compliance
Lightning Source LLC
Chambersburg PA
CBHW021947160426
43195CB00011B/1250